Acclaim for *Monar*

"Cephs rule! *Monarchs of the Sea*, like its pr[...]
ing, smart, and weird in the very coolest [...]
more fun than jetting back in time to prim[...]
really ruled our planet? In these pages, Danna Staaf makes ever[...]
every undersea adventurer's dream come true. It's a fabulous read, with squishy, slimy delight on every page."
—**Sy Montgomery**, *New York Times*–bestselling author of *The Soul of an Octopus*

"This crystal-clear book will open your world to wider horizons and much deeper times. . . . Long before vertebrates evolved anything like higher intelligence, squids and octopuses were on a separate track to versatility, problem-solving, individual recognition, and deceit. Before we can know who we are, we must know who we are here with—and who has come before us."
—**Carl Safina**, *New York Times*-bestselling author of *Beyond Words: What Animals Think and Feel*

"I loved this book. . . . Staaf's approach is short and sweet, well-illustrated and strong on playful narrative."—*Nature*

"It is a treat to come across a writer with such specialized training who is able to turn esoteric knowledge into a page-turning read for all audiences. . . . Staaf captures what is rarely seen outside the ivory tower: scientists talking among themselves with a touch of irreverence. Researchers everywhere will surely relate."—*Science*

"This engaging book may do for early cephalopods what paleontologists did for dinosaurs in the 1960s: spark a public renaissance of appreciation for these magnificent creatures who once ruled the seas."
—**Jennifer Ouellette**, author of *Me, Myself, and Why* and *The Calculus Diaries*

"Intriguing. . . . This in-depth coverage of an often neglected but ecologically vital group will change your view of squid, octopuses, and their relatives."
—*New Scientist*

"A book like [this] is a reminder that in any scientific narrative, there are always two stories at play. There is the history of the subject you're studying, and then there is the history of its discovery."—*New Republic*

"Fiendishly readable."—**The Inquisitive Biologist**

"Fresh and fascinating."—*The Times Literary Supplement*

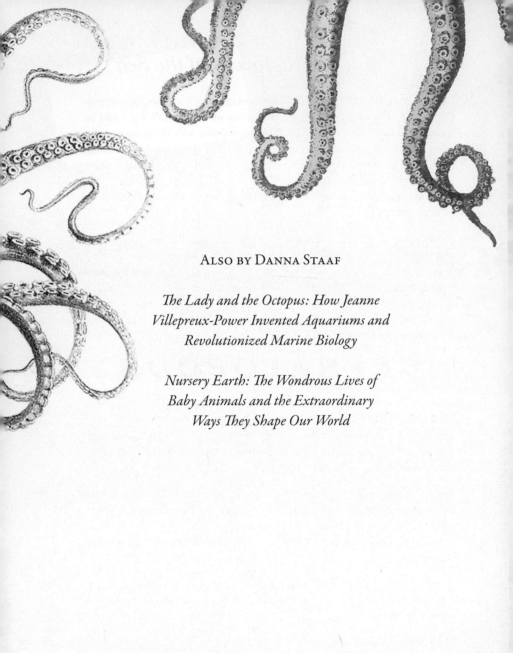

ALSO BY DANNA STAAF

The Lady and the Octopus: How Jeanne Villepreux-Power Invented Aquariums and Revolutionized Marine Biology

Nursery Earth: The Wondrous Lives of Baby Animals and the Extraordinary Ways They Shape Our World

Monarchs
of the Sea

The Extraordinary
500-Million-Year History of
CEPHALOPODS

DANNA STAAF

THE EXPERIMENT

NEW YORK

The Experiment, LLC
220 East 23rd Street, Suite 600
New York, NY 10010-4658
theexperimentpublishing.com

Library of Congress Cataloging-in-Publication Data
Names: Staaf, Danna, author.
Title: Monarchs of the sea : the extraordinary 500-million-year history of
    cephalopods / Danna Staaf.
Description: New York, NY : The Experiment, [2020] | Includes index.
Identifiers: LCCN 2020027541 (print) | LCCN 2020027542 (ebook) | ISBN
    9781615197408 (paperback) | ISBN 9781615197415 (ebook)
Subjects: LCSH: Cephalopoda. | Cephalopoda--Evolution.
Classification: LCC QL430.2 .S728 2020 (print) | LCC QL430.2 (ebook) |
    DDC 594/.5--dc23
LC record available at https://lccn.loc.gov/2020027541
LC ebook record available at https://lccn.loc.gov/2020027542

ISBN 978-1-61519-740-8
Ebook ISBN 978-1-61519-741-5

Cover and text design by Beth Bugler
Author photograph by Josh Weaver

Manufactured in the United States of America

First printing September 2020
10 9 8 7 6 5 4 3 2

For Anton,
the microconch
to my macroconch

Cephalopods, with the unusual means at their disposal, could have become the monarchs of the sea. And they were so, in fact.

—Jacques-Yves Cousteau & Philippe Diolé,
Octopus and Squid: The Soft Intelligence

Contents

Introduction

Why Squid?

Long before humanity was even a twinkle in the eye of the first mammal, our planet was ruled by strange and fearsome creatures. Some grew to monstrous size, the largest animals the earth had ever seen. During their 400 million years of glory they diversified to fill every niche, from voracious predator to placid grazer — and then a global cataclysm almost completely wiped them out. Only a humble few of their descendants survive to keep us company today.

Of course, I'm talking about cephalopods.

I could just as easily have been describing dinosaurs, except for one small hint: the stupendous length of the animals' tenure. Dinosaurs weren't around for nearly as long as cephalopods. Nevertheless, most people know a bit about dinosaurs, while they've never heard of cephalopods. (The accent is on the first syllable: SEF-ah-lo-pod. Some folks in the UK and Europe honor the word's ancient Greek heritage with a hard "C": KEF-ah-lo-pod.) Even those who are familiar with these curious creatures usually know only the living ones — squid and octopuses — not their long-extinct ancestors. I myself was in that camp for quite some time.

I met my first cephalopod on a family road trip when I was ten years old. At the Monterey Bay Aquarium in California, I stood mesmerized by the rippling skin, the undulating arms, and the intimate eyes of a giant Pacific octopus. Shortly after returning home, with my father's patient support, I procured a secondhand saltwater aquarium and became known at school as "the girl with the pet octopus."

I devoured all the information I could find about these amazing animals. In the 1990s, that meant checking out books from the library on sea life and poring over the one or two pages that mentioned cephalopods. I discovered only one exclusively cephalopodic book—*Octopus and Squid: The Soft Intelligence*, by Jacques-Yves Cousteau and Philippe Diole.[1] It was in this book that I first encountered mention of cephalopods as long-ago "monarchs of the sea."

Questions hurried on the heels of this new information. When did octopuses and squid rule the seas? What did their kingdom look like? And why wasn't it around anymore? Cousteau, however, left such pursuit aside and returned to the ever-entrancing study of living cephalopods. So I did the same.

I learned to scuba dive (thanks again to my father, who learned alongside me so I'd have a buddy) and took every available marine biology class. Eleven years after my first visit, I returned to Monterey, this time as a graduate student at the Hopkins Marine Station of Stanford University. Though few people other than marine biologists have heard of this marine laboratory—the second-oldest in the United States—it shares a fence and a friendly professional relationship with the famous Monterey Bay Aquarium.

At Hopkins I spent six years studying the reproductive habits of Humboldt squid. I learned how to drive a boat and cast a net, how to fish with rod and reel in California and with 300 feet (100 m) of hand line in Mexico. I learned how to slice up a piece of squid skin thinner than paper with a glass knife and how to write a computer program that eats up decades of data and spits out a map. I also learned that while I never got tired of explaining the latest squid science to anyone who asked and many who didn't, I often got tired of producing the science myself. After six years, I left Monterey with a doctoral degree and a conviction that I was better suited to science communication than science generation.

In the intervening time, several wonderful cephalopod books had been published,[2] but none about the creatures' heyday as monarchs of the sea. Whenever I sought answers to the questions I'd carried since childhood about their ancient kingdom, I found myself once again limited to one or two pages, this time in books about prehistoric life. And that generally meant books about dinosaurs. The classic dinosaur book hurries through an account of how life evolved in the ocean, diversified into some interesting forms, then finally made its way onto land, where the story *really* gets started.

This bias is quite understandable. Everyone loves dinosaurs, from toddlers playing with plastic *Triceratops* to adults enjoying the *Jurassic Park* franchise, and I am no exception. In one of my first school memories, my second-grade classmates and I read a book of poems called *Tyrannosaurus Was a Beast*, then received the thrilling assignment of choosing a poem to memorize.[3] I picked *Diplodocus*, and one less than thrilling stanza remains indelibly etched: "Diplodocus plodded along long ago, Diplodocus plodded along."

Dinosaur love is so entrenched in our culture (especially our childhood culture) that it's hard to believe it wasn't always this way. But in fact, all through the first half of the twentieth century, dinosaurs were seen as slow, stupid, and boring—not just by the public, but by the very scientists who studied them. Then, in the late 1960s, the legendary Yale paleontologist John Ostrom discovered *Deinonychus* and described it as quick, active, and energetic, in blatant contradiction to established wisdom.[4] Ostrom's student

Bob Bakker, who was equally quick, active, and energetic, and furthermore gifted as both a speaker and an artist, became the champion of what he dubbed the "dinosaur renaissance."[5] The new view of dinosaurs gathered momentum through the 1970s and '80s, though the old "plodding" perspective still showed through from time to time, as in the *Diplodocus* rhyme.

It was Ostrom who taught us that modern birds are surviving dinosaurs, obliging us to refer to ancient forms like *Tyrannosaurus* and *Triceratops* most accurately as "non-avian dinosaurs." It was Bakker who showed us warm-blooded non-avian dinosaurs with complex social behaviors, which were then portrayed in *Jurassic Park* films. That dinosaurs are legitimately enthralling is not a perspective I would dream of quarreling with.

However.

The fossil record of cephalopods goes much further back — 500 million years to the dinosaurs' paltry 230 million. The fossil record of cephalopods illuminates the single most dramatic extinction in Earth's history (yes, more dramatic than the meteor impact that ended the Cretaceous). The fossil record of cephalopods gives Earth some of its most bizarre and beautiful rocks, which humans have interpreted as everything from snakes to pagodas to buffalo. And because of the remarkable way a single cephalopod fossil encapsulates the living animal's history, from embryo to adult, their fossil record may help to explain some of evolution's most recalcitrant puzzles.

What's more, ancient cephalopods share many appealing features with dinosaurs. Massive size is one: the longest fossil cephalopod shells reach 20 feet (6 m), comparable to the height (though not the gobsmacking length) of the biggest dinosaurs. In life, these cephalopods could have borne tentacles that extended their length by several meters. And though cephalopods ruled the seas long before the dinosaurs' ancestors even crawled onto land, the final extinctions of these two great groups were strangely synchronous.

Here is my wonderful secret: although library bookshelves may not be packed with accessible information about ancient cephalopods, the esoteric journals of academia are. Every year—every month, it seems!—cephalopod paleontologists publish new discoveries and new interpretations in the arcane pages of *Acta Palaeontologica* and *Lethaia*. Some of these scientists I met during my years as a student, and they've kindly guided me to other leading lights in the field. I'll be quoting from my interviews with these luminaries throughout the book. From Japan to Germany, from the Falkland Islands to Salt Lake City, researchers are pouring their passion backward in time, striving to understand an ancient watery world. There's never been a better time for cephalopods to enjoy their own renaissance.

Unfortunately, the very name "cephalopod" is a major hindrance. It's not as catchy as "dinosaur," which can be translated as "terrible lizard" even by children with no formal Latin education. What does "cephalopod" even mean? Let's use squid as our entrée (if you'll pardon the gastronomic pun) to the weird and wonderful world of the "head-footed."

Monarchs of the Sea

The World of the Head-Footed

Jet-propelled and flight-capable, iridescent and elastic, squid are a true marvel of nature. They're fast: they can swim twice as fast as an Olympic champion, shoot their tentacles out in less time than it takes you to blink, and alter their appearance at literally the speed of thought. They're flashy: some grow luminous lures at the ends of their arms, others squirt self-portraits in ink, and their skin creates any color from vivid red to iridescent blue.

Yet most humans don't know a lot about squid. We tend to view them through the lens of either mythology or gastronomy, limiting these wondrous creatures to the role of terrifying Kraken or tasty calamari.

If you're in the first camp, kept up at night by visions of Jules Verne's *Twenty Thousand Leagues Under the Sea* or Peter Benchley's *Beast*, then relax and get some sleep. Although giant squid can grow to almost 40 feet, 25 is far more common, and most of that length is in their slender, stretchy tentacles. What's more, there are no reliable accounts of any squid, no matter how giant, endangering boats or killing humans.

On the other hand, if the only thing you know about squid is whether you prefer them fried in rings or raw with rice, you're not alone.

Swimming Protein Bars

Pretty much every animal that encounters a squid tries to eat it. Even bears and wolves have been known to take advantage of the occasional beach stranding.

The poor things are born delicious. Squid produce huge numbers of offspring—from hundreds in some species to hundreds of thousands in others—nearly all of which are eaten before they grow up. When they hatch from their eggs, they're smaller than your fingernail, so Baby's First Predators are small too: fish larvae and aquatic worms.

But squid grow quickly, and in a matter of days or weeks those that survive turn the tables. Growing fat on their onetime adversaries, squid start to attract larger predators: seals and seabirds, sharks and whales. The sheer scale of squid consumption is mind-boggling. Scientists once pumped the stomachs of sixty elephant seals from South Georgia and found 96.2 percent squid by weight.[1] They calculated that the island's elephant seal population scarfs down at least 2.5 million tons (2.3 million metric tons) of squid and octopuses every year.[2] Meanwhile, a single sperm whale can eat 700 to 800 squid every *day*.

Rough luck for the squid. However, it earns them the dubious honor of being named "ecological keystones." They start so small, grow so fast, and get so big that they provide abundant food for marine predators of every size—a "one-prey-fits-all" solution. Thus, many species of squid act as biological conveyor belts, moving energy from tiny plankton up to apex predators. Including humans.

Squid and their octopus and cuttlefish relatives have been caught and eaten by humans for probably as long as humans have

FIGURE 1.1   Humboldt squid can grow up to 6½ feet (2 m) long
and spawn millions of eggs. Their size and abundance
support the world's largest invertebrate fishery.
Carrie Vonderhaar, Ocean Futures Society

lived anywhere near the sea. But the past few decades have seen
an unprecedented boom in commercial squid fisheries, spurred by
the realization of just how many squid were already being eaten
by nonhumans. To give some context, the world's largest fishery
targets Peruvian anchoveta, a small fish that's primarily processed
into fish meal and fish oil. More than 3.3 million tons (3 million
metric tons) of anchoveta were caught by humans in 2014.[3] Compare that with the more than 2.2 million tons (2 million metric
tons) of squid caught by elephant seals . . . from one island. Back in
the 1980s, scientists calculated that the mass of cephalopods eaten
by whales, seals, and seabirds outweighed the total marine catch
(fish, cephalopods, and everything) of all fishing fleets around
the world.[4]

For those familiar with the foibles of humanity, it should be no
surprise that this information caused us to start hauling a lot more
squid out of the ocean.

Most of the world's biggest fisheries target fish, like tuna, cod, and herring. But some fisheries go for other kinds of marine creatures, and the largest of these is for Humboldt squid off the coast of Central and South America. More than a million tons or metric tons of Humboldt squid were caught in 2014, far more than any other non-fish, such as shrimp, lobster, or abalone. That's an explosive growth from the fishery's recent birth in 1965, when it brought in only 110 tons (100 metric tons). (The comedian Chris Rock and the author J. K. Rowling were born in that same year; a comparison of their success with that of the Humboldt squid fishery is left as an exercise for the reader.) The eagerness to catch Humboldt squid is representative of a broad increase in commercial fishing's attention to many kinds of squid over the past few decades. The world's hunger for protein grows along with its population, and when it became clear that many long-standing fisheries (such as the aforementioned tuna, cod, and herring) were already harvested at or above capacity, people began to turn to squid.

Squid hold such a powerful appeal for human and nonhuman predators alike because they lack substantial hard parts — no bones or shells to speak of. Admittedly the ocean is full of other boneless, unshelled animals, like jellyfish, but most of them are gelatinous, containing up to 95 percent water. Squid, on the other hand, are solid muscle and make a far more nourishing meal.[5]

If you've ever eaten squid, chances are you've eaten its mantle — the thick tube that forms the main part of a squid's body. In a living squid, one end of the mantle is sealed closed and adorned with two flexible fins that can flap like wings. The other end is open to the sea. Here the squid sucks water in and squeezes it out through a siphon, creating a jet stream that propels the animal through the sea and even into the air. (Not all squid can fly, but the ones that do can travel up to 165 airborne feet/50 m before splashing down.)

Squid mantles contain the world's largest nerve cells, which squid have used for millennia to jet away from predators and which

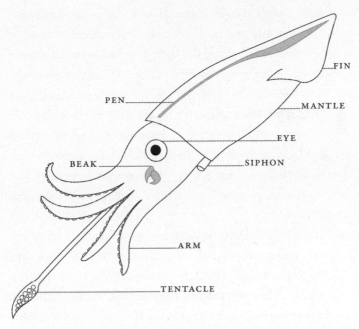

FIN

PEN

MANTLE

EYE

BEAK

SIPHON

ARM

TENTACLE

FIGURE 1.2 A squid's body seems bizarre to human eyes — arms attached to the head, fins at the opposite end, not to mention the suction cups and nozzle-like siphon. But it works for the squid.
C. A. Clark

scientists have used for eight decades to elucidate pretty much all of modern neuroscience. In the 1930s, the English scientists Alan Hodgkin and Andrew Huxley figured out how to stick needles into these enormous cells (fifty times wider than the widest nerve cell of a mammal) and measured for the first time the electrical signals all nerves use to communicate.[6] With the Australian scientist John Eccles, who had worked on complementary research, Hodgkin and Huxley were awarded a Nobel Prize in 1963. Since then, researchers have developed technology to measure electrical signals from much smaller nerve cells, like the ones in our own brains. It was squid that showed them what to look for.

Squid have brains too, but instead of a double-hemisphered lump, they come in three parts: one optic lobe behind the left eye, another optic lobe behind the right eye, and between them a strangely shaped doughnut of nervous tissue. Through the doughnut hole runs the squid's esophagus. This is the most direct route from the mouth into the mantle, where the stomach and other organs lie, but as you might imagine, swallowing through your brain can be risky. The squid must ensure that each bite is small enough to pass and devoid of sharp bones. The arms and tentacles that surround its mouth provide key assistance during this careful culinary operation.

Squid and all their relatives share the anatomical feature of appendages sprouting directly from the cranium; this is the source of the name "cephalopod," Greek for "head-foot." An octopus has eight arms only, while a squid has eight arms and two tentacles. The distinction between arms and tentacles is easy to remember from the words themselves: *arms* are shorter than *tentacles*. Also, we humans have arms, and cephalopod and human arms share the reassuring feature of always staying the same length. By contrast, cephalopod tentacles are disturbingly stretchy and get scrunched into hidden pockets when not in use.

Cephalopod arms do differ somewhat from our own in being fully lined with suckers. The elastic length of a tentacle bears no suckers at all; suckers would interfere with stretching. Instead, the tip of the tentacle broadens into a sucker-covered *tentacular club* for gripping prey.

Imagine that the squid sees a fish. In a few hundredths of a second, the two tentacles shoot out. Their flattened clubs suction onto the fish and yank it back toward the squid's head. Now the eight sucker-lined arms embrace the prey, while a hawklike beak severs its spinal cord. The squid proceeds to take one bite at a time, swallowing with the help of a rasping tongue called a *radula* and hoping not to skewer its brain.

Squid are pretty good at avoiding sharp bones when they eat fish, and like every other predator they're also happy to avoid bones by eating... other squid. Yep, most squid will eagerly take any opportunity to chow down on their fellows, to the point that in one particular deep-sea species, cannibalism reportedly accounts for 42 percent of its food.[7]

Given their popularity as a menu item, one has to wonder why squid didn't take the sensible precaution of their equally muscular but better-armored cousins, clams and mussels. Why not protect their luscious flesh with a hard shell?

Actually, squid did.

The Squid's Family Tree

Squid are close cousins of octopuses and cuttlefish, more distant cousins of the pearly nautilus, and still more distantly related to snails and clams. Of course, now that I've used the word "octopuses" I find myself obliged to address the perennial question: "octopuses" or "octopi"? Or, heaven help us, "octopodes"?

Whichever you like best. Seriously. Despite what you may have heard, "octopus" is neither ancient Greek nor Latin. Aristotle called the animal *polypous* for its "many feet." The ancient Romans borrowed this word and latinized the spelling to *polypus*. It was much later that Renaissance scientists coined and popularized the word "octopus," using Greek roots for "eight" and "foot" but Latin spelling.

If the word had actually been Greek, it would have been spelled *octopous* and pluralized *octopodes*. If translated into Latin, it might have become *octopes* and pluralized *octopedes*, but more likely the ancient Romans would have simply borrowed the Greek word — as they did with *polypus*. Various ancient Romans pluralized *polypus* in various ways; those who perhaps wished to appear erudite used the Greek plural *polypodes*, while others favored a Latin ending and pluralized it *polypi*.

The latter is a tactic we English speakers emulate when we welcome "octopus" into our own language and pluralize it "octopuses," as I've chosen to do.[8]

We can have pretty much the same debate about "nautilus." The ancient Greek word *nautilos* meant "sailor," and again it was latter-day scientists who latinized the word to name the animal "nautilus." For some reason "nautili" holds a much weaker popular appeal than "octopi," and you nearly always see the plural "nautiluses."

The terms "squid" and "cuttlefish" are less tendentious, and are most commonly used for both singular and plural. Occasionally it is useful to distinguish between multiple "squid" of a single species and multiple "squids" of different species, but it hasn't seemed necessary in this book.

Now that that's taken care of, let's have a look at the whole lot of squishy, slimy, oozy animals called "mollusks," which include cephalopods and all their cousins. ("Mollusk" comes from the Latin word for "soft"; one has to wonder if ancient Romans used the same word to denote "squishy" or "slimy.") Molluscan bodies are broadly divided into two parts: a muscular foot and a shell-secreting mantle. Like different poets improvising with the same poetry form, different mollusks have adapted the same body plan for various lifestyles. Snails ooze along on their foot and carry a coiled shell on their back; clams dig into the mud with their foot and hide in a hinged shell; squid divide their foot into arms and tentacles and repurpose the mantle for jet propulsion, shrugging off its shell-producing capabilities.

However, the very first cephalopods were literally defined by their shells. They evolved from creatures that weren't exactly snails but looked and behaved a lot like snails, crawling along the ocean floor under weighty homes. Then some of these not-snails did a curious thing. While all the other animals of the time continued

FIGURE 1.3 The cut shell of a modern
nautilus displays a logarithmic spiral.
Wikimedia Commons user Chris 73

to burrow in or grovel on or cruise over the seafloor, the far-distant ancestors of today's squid filled their shells with gas and floated up through the water.

They were slow swimmers, but they had no need for speed. They could drift over the bottom-dwelling buffet like deadly dirigibles, selecting their prey at leisure. Two hundred and fifty million years before the first dinosaur, cephalopods became the planet's primary predators, and it was all thanks to their buoyant shells.

Over time their lineage separated into three main branches: nautiloids, coleoids, and ammonoids. The "-oid" suffix is common in zoological nomenclature. It sounds a bit goofy, but it's important because it tells you that I'm talking about a whole group of animals. "Nautiloids," for example, refers to every species there ever was on this family branch, including hundreds of extinct ones as well as the few living nautiluses — today's only shelled cephalopods.

Unless you've visited one of the few aquariums that display nautiluses, you've probably never seen one alive. It's more likely that you've seen a nautilus shell—maybe whole and tiger-striped, maybe polished to mother-of-pearl, maybe sliced in half to show its spectacular spiral. Nautilus shells are so beautiful that humans can't seem to leave them in the water. After decades of increasing concern that nautiluses may not survive our intense and unregulated urge to collect them, in 2016 an international conference finally agreed to monitor and control the trade in nautilus shells. Scientists hope that this protection by the Convention on International Trade in Endangered Species (the same treaty that protects high-profile species like lions and elephants) will safeguard the ancient nautiloid lineage against untimely pruning.

This lineage is legitimately ancient, although neither the nautilus species we know today nor the larger nautiloid group goes all the way back to the dawn of cephalopods. There's been some confusion on this point, because modern nautilus shells do look superficially similar to those of ancient cephalopods, and nautiluses are sometimes called living fossils. For a long time, even professional paleontologists used "nautiloid" as a catchall term for any cephalopod that didn't obviously belong to any other group, including the very first floating snails.

However, despite shell similarities with early fossils, today's nautiluses have evolved their own contemporary peculiarities. Their heads bear not eight, not ten, but anywhere between sixty and ninety tentacles. The number can vary even within a single species. To add to the confusion, nautilus tentacles are not elastic and have no suckers. Instead, each tentacle consists of a protective sheath and a long thin sticky part that can be extended or withdrawn. What's more, at some point in the nautilus's evolution the two uppermost tentacles seem to have enlarged and fused to form a protective hood over the animal.[9] It's enough to make you throw up your hands and study snails instead, with one simple foot each.

But let's not forget that cephalopod appendages are really modified feet. If you watch baby squid and nautiluses develop in their eggs, it's easy to see the connection. In the same way that human embryos retain tails from our evolutionary history, cephalopod embryos show off their molluscan heritage with a single foot that eventually differentiates into ten arm buds. Even nautilus embryos go through the ten-arm stage before these buds divide further.[10] This indicates that early cephalopods, and probably even early nautiloids, all had ten arms; it was some unknown selective pressure later in nautiloid evolution that led to an eventual profusion of tentacles.

Paleontologists are now quite clear that for the first few million years of cephalopod history, there were no real nautiloids, just a variety of evolutionary experiments with tongue-twisting names like plectronocerids, ellesmocerids, and orthocerids. (Feel free to forget those names immediately.) The term "nautiloid" has been reclaimed for the branch that would eventually sprout modern nautiluses, a branch that arose after a lot of those early oddballs had already died out. So while the nautiloid lineage is certainly venerable, it doesn't go *all* the way back.

In fact, nautiloids are only a little bit older — by geologic standards, anyway — than the first coleoids and ammonoids. These latter two groups seem to have evolved in response to competition from and predation by the Paleozoic nouveau riche: fish. A few weird little fishy-looking things had been squiggling around for a hundred million years or so without making much of a splash, but with the evolution of real fish, the marine game changed. Fish could grow several times bigger than the biggest cephalopods, swim significantly faster, and break shells with their jaws. As these bony upstarts began to throw their weight around the prehistoric seas, nautiloids faded into the background while coleoids and ammonoids thrived.

The word "coleoid" comes from the Greek word for "scabbard." A scabbard covers a sword, and the bodies of coleoids cover their shells (or lack thereof). Coleoids include all the modern non-nautilus cephalopods — octopuses, squid, cuttlefish, and more — as well as numerous fossils. Although octopuses, in their intense squishiness, retain barely the vestige of a shell, squid and cuttlefish both have ghostly shells. A thin internal rod called a *gladius* stiffens the squid's body and gives its muscles something to work against. A cuttlefish, which looks almost exactly like a squid on the outside, has a more complex calcified internal structure called a *cuttlebone*. You may have seen one hanging in a birdcage, its calcium serving as a dietary supplement for the avian inhabitant.

Ditching the safety of the shell may seem to have been a monumentally stupid move on the part of evolution. On the other hand, it freed these animals to develop the array of remarkable adaptations they're known for today.[11] Aquarists learned long ago that the only limit on an octopus's ability to squeeze through holes is the size of its beak — its single unsquishable body part. More than one proud octopus owner has gone to check on her pet, only to commence a panicked search for the absent animal in aquarium filters and plumbing and even on the floor around the tank. Octopuses can survive for a short time out of water but will eventually suffocate, so the hunt ends sometimes in relief and sometimes in tragedy. Forewarned by aquarium magazines, I used a liberal application of plastic wrap and duct tape to escape-proof my tank, and managed to avoid disaster during my own octopus-keeping years.

The great escape of one captive octopus from a New Zealand aquarium in 2016 had the most triumphant ending possible: successful return to the sea. "Inky" found a gap in his enclosure and made tracks across the floor to a drain, then slid down the narrow pipe directly into the ocean. Prison breaks are hardly the only application of the octopus's Houdini skills — in his new wild life, Inky will ooze under rocks in pursuit of prey and frustrate predators by disappearing into crevices.

Prey and predators alike are also regularly flummoxed by coleoid skin, the most complex camouflage system in nature. The word "chameleonic" should really be "cephalopodic" to reflect the fact that squid, octopuses, and cuttlefish are much better than any reptile at instantly matching their environment. The chameleon's skin-changing system relies on hormones, which have to be manufactured in the brain and then distributed around the body in the bloodstream. The cephalopod's system is under direct nervous control. Each spot of color (of which there can be more than two hundred in a square millimeter of skin) is controlled by tiny nerves that connect straight back to the brain. "Rapid" color change in chameleons takes a couple of minutes;[12] changes in squid skin have been clocked at up to four per second.[13]

Modern coleoids are so amazing that it's hard to imagine them as anything other than an unmitigated success throughout their evolutionary history. But it's unclear how abundant and diverse the coleoids really were, because their fossil record is limited by evolution's modification, reduction, and in some species total deletion of the shell. A soft body is far less likely to fossilize than a hard shell. Although people have been writing about shelled cephalopod fossils since ancient times, the first fossil octopus wasn't described until 1883. What we can glean of the coleoids' evolutionary history from such limited evidence, though, suggests that for a long time they played second fiddle to the ammonoids.

From their first appearance through many geologic periods after, ammonoids were the cephalopods' big success story. Although none survived to grace our modern seas, ammonoid shells are some of the world's most common, and most lovely, fossils. Curled into spirals that some people called snakestones, they were eventually named for the ancient god Ammon, who wore ram's horns.[14]

Because the fossils they left behind are coiled external shells like those of the modern nautilus, living ammonoids were initially

FIGURE 1.4 This "nautilus-style" reconstruction of
fossil ammonoids with thick hoods and dozens of tentacles
was created around 1916 by Heinrich Harder.
Photo from Tiergarten, Berlin, by C. A. Clark.

assumed to share the nautilus's soft-body features as well. Artistic reconstructions used to include de rigueur a thick fleshy hood atop ammonoid heads and an embarrassing wealth of tentacles. However, further studies of ancestral relationships indicate that ammonoids were more closely related to coleoids, and this perspective is reflected in all the latest artistic reconstructions.

The smallest adult ammonoid coils were mere centimeters in diameter; the largest reached 6½ feet (2 m). You could crawl into a shell that big, as long as the ammonoid wasn't in residence. Some coils were so loose that water could flow between the loops; others were so tight that they grew over themselves. Some were thin and some were fat; some were simple and some were fancy.

Ammonoids were so abundant and evolved so quickly that paleontologists use them to determine the age of rocks. A particular species of ammonoid can often be tied to a particular slice of geologic time. For example, anywhere you find the pretty spiral of

FIGURE 1.5 This "squid-style" reconstruction was published
in 2015, with each ammonoid sporting eight arms,
two tentacles, and a large muscular funnel.
Andrey Atuchin, in A. A. Mironenko, "The soft-tissue
attachment scars in Late Jurassic ammonites from Central Russia,"
Acta Palaeontologica Polonica 60, no. 4 (2015): 981–1000.

a *Dactylioceras athleticum*, you can be sure that the surrounding
rock is between 182 million and 175.6 million years old.

Does that seem . . . not especially useful? Let's take a step back,
then, and consider the simple little task of understanding Earth's
history.

Rock Clocks

Fossils are usually made from bodies, but can also be created by
footprints or feces. The world is so full of bodies and their by-
products that you might expect to be tripping over fossils at every
turn, but in fact fossils are wholly improbable.

Think about the last dead body you saw. The spider curled up
on the windowsill will probably dry out and disintegrate. The rac-
coon on the highway shoulder will be devoured by scavengers, and
any leftover bones will be broken, weathered, scattered.

These examples are representative of what happens to most
animals when they die—at least, those that aren't completely

FIGURE 1.6 A collection of fossilized
Dactylioceras ammonoids from the Lower Jurassic.
Istvan Takacs

consumed and digested. Animals thus fated do not form fossils. Certain human cultural practices notwithstanding, it's the rare body that is buried whole. Good fossils require unusual circumstances. A volcanic eruption, a tar pit, a mudslide. Amber, the fossilized tree juice that famously (and fictionally) preserved dinosaur DNA in Jurassic Park, even once preserved an ammonite shell, perhaps because a tree dripped its resin on the beach.[15] Unfortunately, the ammonite's soft parts had already been eaten or rotted away, so the amber offered no new anatomical insights.

Even those fossils that do get formed are invisible until unburied. The simplest unburial consists of many long years of erosion, with the occasional landslide or earthquake to hurry things along. Human excavations and explosions can also reveal fossils, and yes, we are pretty good at digging holes and blowing stuff up, but in the grand geologic context these are just a few dents. The majority of the world's fossils lie deep beneath our feet, below the seafloor, never to be seen.

Given that most life never fossilizes and most fossils are never found, geologists as well as paleontologists show an understandable enthusiasm for those organisms like ammonoids that *do* show up abundantly in the fossil record. If they are abundant and diverse enough, they can tell time.

You see, people have long been aware that Earth's rocks come in layers. The idea that the top layers are the youngest, and the bottom layers the oldest, goes back at least to the sixteenth century. But because nobody at the time or for the next four centuries had any real clue how old the whole Earth was, dividing Earth's history into rock layers was nothing like dividing a day into hours and minutes. Rather, rock layers were identified by their content — chalk, coal, or limestone — and named after the places where scientists first identified them, such as the "Permian" from Perm, Russia, and the "Devonian" from Devon, England (both defined in the 1840s). At first, rock layers in different places had seemed quite different from each other. Fossils provided the key to broadening and standardizing these names for use around the world.

Scientists noticed that the same fossils or combinations of fossils, including many ammonoids, often showed up in different places. They could be used like fingerprints to identify a given unit of geologic time. By the mid-nineteenth century, thanks to quite a few hardworking geologists and many more long-dead cephalopods, Earth had been granted a geologic timescale. A hundred years later, thanks to radiometric dating, we were finally able to attach precise numbers to it.

To understand radiometric dating, first consider that rocks are made of elements: carbon, oxygen, calcium, and so forth. Some of these elements (most notably, uranium) come in lighter and heavier forms. Some of the heavier forms are unstable and tend to spit out tiny bits of themselves until they reach a stable, lighter form. For each form, we can figure out the rate at which the spitting happens. Then we (and I use "we" in the sense of "other members of my species who are far more skilled in this technique than I") take a piece of rock, measure the relative amounts of unstable and stable elemental forms, and use those to calculate how long the unstable forms have been spitting themselves stable. That tells us when the rock was first formed, and—finally!—how old it is.

To a geologist, the word "eon" is more specific than "a really long time." Eons are the largest units into which Earth's entire 4-billion-year history is divided. Eons are then subdivided into eras, and eras into periods. Periods are the units of geologic time you're most likely to have heard of, with names like Jurassic and Cretaceous (as well as the Permian and Devonian mentioned earlier).

For our purposes here, we need consider only one eon, which is the eon we still live in today: the eon of "visible life" or Phanerozoic, which is just half a billion years long. This eon contains three eras: "old life" (Paleozoic), "middle life" (Mesozoic), and "new life" (Cenozoic), each with its own constituent periods. With the scientist's passion for precision, the periods have been subdivided into epochs and ages, but we need not concern ourselves with those—except to appreciate that a great deal of this precision is owed to ammonoids.

Ammonoids serve as ideal geologic timestamps. Their uncannily speedy evolution means there's practically a new species every geological "minute." Their fossil abundance means you can find the same "fingerprint" in many kinds of rocks, in many places around the world. The only regrettable aspect is that seeing ammonoids as rock clocks for a long time made it hard to see them as anything else.

"Ammonites were fossils rather than fossil organisms: scientists had discussed the way one species gave rise to another, and to the distribution of each species across the globe. But what ammonites actually did when alive was very, very vague," says Neale Monks,[16] who wrote the book about ammonoids, despite the title *Ammonites*.[17] That's because when people discuss ammonoids,

even if those people are professional paleontologists, they tend
to use the more familiar and casual word "ammonite," though as
Monks notes in the preface to his book, "strictly speaking 'ammonite' is used for the single suborder Ammonitina with the Ammonoidea."[18] I hope I may be forgiven for instead using the more
obscure and formal "ammonoid," on the grounds of producing a
pleasant parallelism with "nautiloid" and "coleoid."

Though Monks did a stint as a professional paleontologist, he
started out as a keen hobby aquarist and obtained his first university degree in zoology. Accustomed to thinking of the animals in
his tanks as living, breathing, pooping organisms, he was a bit star-
tled to enter graduate school in paleontology and discover that
ammonoids had long been treated primarily as convenient rocks.

Fascinated by questions of living ammonoid biology, Monks
found a kindred spirit in Philip Palmer, a curator of fossil mollusks at the Natural History Museum in London. Eventually the
two decided to turn their hours of discussion into a book. With
the publication of *Ammonites* in 2002, Monks and Palmer pro-
claimed: These rocks were once alive. Here's where the animals
might have lived, how they might have moved, what they might
have eaten.

But Monks is also well aware of the limitations intrinsic to this
kind of speculation. In a 2016 article titled "Ammonite Wars," he
discusses why it's been so difficult to make biological sense of their
fossils: "Vertebrate bones have an intimate relationship with the
muscles connected to them. Looking at the skeleton of a dinosaur
or mammoth tells you a great deal about how the animal was put
together and what it looked like in life. By contrast, the shell of
an ammonite is mute. Apart from a few vague muscle attachment
scars, there's little to be gleaned about the size and shape of the
ammonite animal's soft body parts."[19]

What ammonoid fossils lack in soft-body details, though, they
make up abundantly in tales of birth, growth, and maturity. When

it comes to the development of the organism over its lifetime, ammonoid shells speak volumes. And development is emerging as
one of the keys — perhaps *the* key — to understanding evolution.

A New Poster Child for Evolution

There's a lot we do know about evolution. We know that all life on
the planet is related, and we can trace our relationships in DNA.
We know that through natural selection each species adapts to its
own niche and that, every so often, mass extinctions empty vast
swaths of niches, presenting those species that survive with new
opportunities to adapt. We know that both evolution and extinction can happen very quickly — we've seen insects and bacteria evolve resistance to our attempts to kill them, and we've seen
dodos and sea cows fail to do so.

But we still have a lot to learn.

One of the great projects in the study of evolution is to understand the origins of novelty. Where do we get new forms, new
patterns, and new habits on the scale necessary to produce the
staggering diversity of life around us? Natural selection, that brilliant Darwinian brain wave, could be anthropomorphized as a
sculptor. Whence the clay?

Answers have begun to arise from a field of science known by
the nickname "evo-devo." It sounds like the name of an indie rock
band, but it's shorthand for "evolution and development," and it's
deeply rooted in genetics.[20]

Our DNA, as it turns out, is not a linear, step-by-step assem-
bly manual. It's more like a wiring diagram, a network of inter-
acting connections. Each organism starts out at the beginning of
its life looking roughly the same — a single cell ready to grow —
and many of the genes inside that cell are roughly the same across
the animal kingdom. Higher-level controls in each cell determine
which steps of organismal construction are ignored and which

are followed, as well as when they are followed, in what order, and how many times. Slight changes in these high-level controls can lead to extreme novelty: a new number of limbs, a new body shape, a new kind of reptile scale that's actually a feather.

Our current and still-limited understanding of evolution is comparable to early scientists' incomplete understanding of the physical changes in our planet. They knew that Earth had altered over its history, but struggled to understand how and why, until the idea of plate tectonics gained traction. Once we understood that the planet's crust was made of floating, interlocking plates, everything made sense—from the shapes of continents to the distribution of marsupials.[21]

It's possible that we're in the midst of a similar breakthrough in evolutionary biology, as the astonishing example of the four-winged dinosaurs suggests.[22]

Dinosaurs are the poster children for evolution and extinction writ large. Most of us are fascinated by these creatures as soon as we're old enough to ask questions: They got *how* big? *Why?* And they all *died*? *How?*—the same questions that grown paleontologists continue to wrestle with, ever gathering new data to update the answers.

Of course, not all of them did die. We know now that birds are simply modern dinosaurs, but out of habit we tend to reserve the word "dinosaur" for the huge ancient creatures that went extinct at the end of the Cretaceous. After all, even if they had feathers, they seem so *different* from today's finches and robins. For one thing, the first flying feathered dinosaurs all seem to have had four wings. There aren't any modern birds with four wings.

Well . . . actually, domestic pigeons can be bred to grow feathers on their legs. Not fuzzy down, but long flight feathers, and along with these feathers their leg bones grow more winglike. The legs are still legs; they can't be used to fly like wings. They do, however, suggest a clear step along the road from four-winged dinosaurs

FIGURE 1.7 The ammonoid *Crioceratites* lived in the
early Cretaceous. Its beautiful fossil preserves a
record of the animal's growth.
Franz Anthony

to two-winged birds. The difference between pigeons with ordinary legs and pigeons with wing-legs is created by control switches
in their DNA that alter the expression of two particular genes.[23]
These genes are found in all birds, indeed in all vertebrates, and
so were most likely present in dinosaurs as well. A simple change
in their expression, begun in the egg, makes a drastic difference in
the grown animal. That's evo-devo.

Studying gene control and expression in embryos to find out
what happens as they grow is becoming a viable research technique in cephalopods too. Eric Edsinger-Gonzales, a biologist
at the Marine Biological Laboratory in Woods Hole, Massachu-
setts, has worked with the embryos of numerous cephalopod species, and he's confident that in the next few years scientists will be
able to genetically alter cephalopods to create "lines" somewhat
like pigeon breeds. I wondered what discoveries might be lurking

on the same scale as wing-legs in pigeons, and Edsinger-Gonzales suggested that the right genetic tweak might cause an octopus to grow ten, instead of eight, appendages. I can imagine that perhaps even a nautilus embryo might be induced to scale back to ten arms, providing a dramatic demonstration of how such shifts could have happened over the course of cephalopod evolution.

Sadly, squishy arms don't fossilize nearly as well as bony legs and wings, and we may never know for sure how many limbs ancient cephalopods had. Still, fossil cephalopods possess two distinct advantages over dinosaurs when it comes to the study of evolution. One: their shells contain a record of their growth all the way back to the egg, so developmental changes—many of which might be driven by DNA control switches—can be traced through a single animal's lifetime. Two: they were waaaaaaay more abundant.

These perks appealed to the American paleontologist Peg Yacobucci, currently a leader in the field of fossil cephalopods, who happily admits, "I was a dinosaur nut as a kid, of course."[24] As a "dorky high school student" in the 1980s, she was fascinated by the exciting new idea that an asteroid impact killed off the dinosaurs. When the eminent paleontologist Jack Sepkoski gave a talk near where her family lived, Yacobucci says, "I had my mother— my poor mother!—take me to the science museum to see this talk. She fell fast asleep."

Sepkoski did not study dinosaurs; he studied mass extinctions. To elucidate patterns in extinctions he used patterns in marine fossils, including cephalopods as well as even smaller and less sexy things, like clams and plankton.[25] All are abundant, diverse creatures with a rich fossil record—something dinosaurs notably lack. For all their abundance on bookshelves and in toy shops, dinosaur fossils in the field are rare.

At the end of Sepkoski's talk, he invited the audience to return the following day, when he would be discussing, instead of extinction, the opposite process: evolution. Yacobucci insisted on another ride from her mother and was even more entranced. She remembers Sepkoski essentially saying, "It's easy to kill stuff off, but how do you explain the rapid appearance of new groups?" She goes on, "He talked about the Cambrian Explosion, and I'd never even heard of this event, and it was even more fascinating to me. It fixed for me my research question, which I still have: where do new species come from?"

Yacobucci went off to college at the University of Chicago, because that's where Sepkoski worked, and she initially studied dinosaurs. But, she says, she quickly realized, "I can't do what I want to do intellectually with dinosaurs. Dinosaurs aren't common enough to do rigorous studies of evolution." For her graduate research, she said to herself, "I need a group; it's got to be just as cool as dinosaurs. It would be great if it lived at the same time as dinosaurs; bonus points if it dies out at the same time as dinosaurs, but with a really rich fossil record."

Within all these constraints, the choice was obvious.

In the following chapters, we'll trace the ups and downs of the evolution of cephalopods, from crawling snails to slow-floating predators to fast-moving escape artists. We'll witness triumph after triumph of the ammonoids, followed by a catastrophic doom that the simple, unassuming nautiloids somehow sail through. We'll watch as the cephalopod shell undergoes endless change: coiling, truncating, growing in knots, and eventually, in the coleoids, internalizing and dissolving. Finally, we'll survey the diversity of modern cephalopods, from giant squid to flamboyant cuttlefish, and ponder what their future might hold.

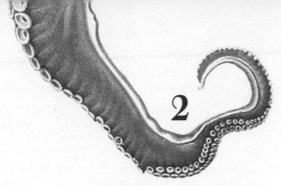

2

Rise of the Empire

Squid and octopuses are so weird it's tempting to call them aliens. The tentacles. The bonelessness. The disturbingly quick skin changes. Even the scientists who study them (perhaps *especially* the scientists who study them) are struck by their otherworldly nature.

When an octopus's complete genetic code was sequenced for the first time in 2015,[1] a leader of the research team joked, "It's the first sequenced genome from something like an alien." Headline writers around the world rejoiced, producing gems such as "Scientists Declare That Octopuses Are Basically Aliens" (Geek.com) and "Octopus Genetic Code Is So Strange It Could Be an ALIEN" (*UK Mirror*).

The particular octopus whose genes were extracted and identified was a sweet little species called the California two-spot octopus, beloved of aquarists because it tends to be tolerant and easygoing. (My second octopus, Rex, was a two-spot.) Scientists found hundreds of genes unique to octopuses, as well as a surprising abundance of a group of genes called protocadherins. As more cephalopod genomes have been sequenced, including bobtail squid and giant squid, all have turned up rich in protocadherins.

Such abundance has been seen in no other animals—except vertebrates.

Vertebrates, that's fish and turtles and cats and dogs and you and me. The first vertebrates evolved around the same time as the first cephalopods, and as soon as they got fishy, vertebrates became a powerful driver of natural selection on cephalopods. Squid and octopuses have even been called "invertebrate fish," with convergent features in body, brain, and behavior.[2] Cephalopod genomes may offer a glimpse into how creatures that were essentially snails evolved to be practically fish.

And yet, as Edsinger-Gonzales, who worked on the octopus study, says, "The genome doesn't look that different from a snail's."[3] He had previously helped to sequence the genome of a kind of marine snail called a *limpet*, so he's well equipped to compare the two. "There are some genes that we only know for octopus right now. But if you go to the snail we worked on before, there's also the same percentage of genes we only know for the snail."

Edsinger-Gonzales doesn't find that boring or discouraging, though. *Au contraire*! With the scientist's irrepressible enthusiasm, he appreciates "how typical the genome looked. I thought that was really exciting . . . You can have the same gene doing six different things in six different animals; you only know when you test it what it's doing in each case. That means there's a lot of interesting work to be done on cephalopods."

It also means that cephalopods are undeniably related to every other form of life on Earth. Their genome proves it—in addition to its similarity to a snail's, it is full of sequences with matching counterparts in humans and fruit flies alike.

Disappointed? Well, cheer up, because there's a chance we can *all* claim an extraterrestrial heritage.

Setting the Stage

Scientists still aren't sure how life originated on our planet, but it's definitely been here for a long, *long* time. The earliest fossil evidence of life is found in rocks 3.7 billion years old.[4] Earth itself is less than a billion years older than that.

These oldest fossils are simple cells, which is both boring and amazing. We ourselves are much bigger than cells, and naturally we're interested in other big things, like trees and seashells and dinosaurs. But small as it is, early evidence of cellular life is quite uncanny. Cells are complicated, with membranes and DNA and numerous other components that all have to play nicely together. Such complex structures didn't spring fully formed from the rocks — they must be the result of evolution's long slow grind acting on simpler precursors, like self-replicating molecules. Such molecules would have to be older than 3.7 billion years. Maybe even as old as Earth itself.

And yet for much of the first billion years of its existence, Earth was too scorchingly hot to support any of the proto-living molecules we can imagine. So did life come to the eventually cooled planet from . . . elsewhere?

In those early days, the young solar system was full of whizzing projectiles, and our spherical home suffered a continuous pounding. Organic compounds, like the acids and sugars that make up DNA, do exist extraterrestrially. They could have been smooshed into Earth in bulk quantities during that tumultuous era. What's more, we know that meteorites can (and did!) bang into Mars, knock off planetary chunks, and send them hurtling Earthward. And young Mars of 4 billion years ago was a much friendlier environment than young Earth, with a pleasant temperature and a nice combination of land and water. A number of scientists believe it is possible — even probable — that life, either complex molecules or actual cells, evolved first on Mars and then seeded Earth.[5]

Thus, the raw material that you and an octopus both use to build your bodies may have originally come from outer space. That's a pretty cool thing to have in common.

However, although cells can grow into all sorts of wondrous bodies today, it took an extraordinarily long time for them to arrive at this ability. While molecules (as far as we know) transitioned to cells in well under a billion years, life then proceeded to remain locked in single cells for the next *3 billion years*. The cells evolved, certainly, and sometimes grew together in colonies, but there was no specialization, no division of labor. And that means there were no animals. An animal body needs, for a start, skin cells to cover it, muscle cells to move it, and sensory cells to see, sniff, and slurp its surroundings.

Why did it take so long for cells to team up the way their component molecules once did, to grow into tissues and organs and *creatures*? Maybe the planet's environment wasn't conducive to it. Maybe a few cells tried, but there wasn't much advantage and they died out without leaving a trace of the experiment. But something changed around 700–800 million years ago, and Earth got to meet its first animal resident—the humble sponge.

Sponges are barely animals. They can't move and they have no sense organs, or indeed organs of any kind. But they're more than a colony—they contain specialized cells, constituting a proper *organization*. And they were the only animal game in town for the next 100 or 200 million years. They slowly built up genetic and cellular diversity, though nothing like what we think of in terms of animal diversity today. That arrived, at last, in the Ediacaran period, 600 million years ago, and animals were off and running.

Well, in a way. They were sitting on the seafloor rather than swimming off it, and stuck in the mud rather than running. Actually, they were nothing at all like animals you'd recognize today. They were tall fronds swaying in the water and oval blobs nestled in the mud. They could get reasonably large—one frond grew to

6½ feet (2 m)—but they had no arms or legs or tentacles, no guts or snouts or eyes.

One strange denizen of the Ediacaran *did* get a bit more complex than the standard of its time. The round fossil *Kimberella* appears to have had a mouth, which left scraping marks on the ground. These fossilized scrapes look an awful lot like the marks left by modern snails as they graze on algae with a tonguelike tool known as a *radula*. You may recall that squid also have a radula, which they put to more ferocious use—that's because squid and snails are both mollusks, and the radula is a molluscan invention. *Kimberella* may be their most distant ancestor.[6]

Then again, some scientists have interpreted *Kimberella* as a jellyfish. With these peculiar Ediacaran fossils, it's kind of hard to be sure.

In order to see familiar animals, like starfish and shrimp and the all-important seashells that will eventually (and literally) give rise to our dear cephalopods, we'll have to jump ahead—50 million years—to the Cambrian Explosion, the evolutionary party that catalyzed Yacobucci's lifelong research question. This grand event was heralded by an unassuming little worm.

The World's Slowest Explosion

Our modern world is full of worms. The word "worm" is almost uselessly vague, as it describes so many different animals. There are the segmented worms, a group that includes both the humble earthworm and the ornate Christmas tree worm. There are the aggressively carnivorous ribbon worms, the usually plant-sucking roundworms (but the parasitic hookworms and heartworms are also types of roundworms) and even slow worms, which are really legless lizards.

And then there are the penis worms. Yes, that is their real name; biologists can get away with that sort of thing because of their classical education. The scientific name of the group is Priapulida, and Priapus was a Greek fertility god whose statues and frescoes commonly sport a leg-sized erection. It's hard to deny a visual similarity to the plump, proboscis-tipped little worms. But there's more than ribald humor going on here—both worm and phallus derive their shapes from the same anatomical structure, a hydrostatic skeleton.

This remarkable boneless skeleton supports and transmits muscular force with fluid alone. It was, in fact, the world's first skeleton. Penis worms today use their hydrostatic skeletons to make burrows with a distinctive shape, and paleontologists have discovered similar shapes in rocks 542 million years old. Like *Kimberella*'s scrape marks that connect it to modern mollusks, these ancient burrows were certainly created, if not by true priapulids, then by something very similar indeed.

The actual burrower has never been found; probably it was too soft to fossilize. The burrows have been given their own scientific name, *Treptichnus pedum*, and the honor of marking the onset of the Cambrian period—which is the first period of the Paleozoic era, which is in turn the first era of the Phanerozoic eon, so *Treptichnus* marks the start of all three. It might seem strange to make such a fuss over a "trace fossil," which is not even an animal but only evidence thereof. But *Treptichnus* is the first fossil that shows a creature moving down into the ground, creating a three-dimensional space with its body. This is a significantly more complex behavior than anything seen in the Ediacaran, and it requires a more complex kind of body—a "modern" animal's body.

Not only penis worms but many of the animal forms we know today are first found preserved in the famed (at least among fossil aficionados) *Cambrian Explosion*.[7] This sudden profusion of life may have been kicked off by changes in the earth's physical environment. Various lines of evidence suggest that animal evolution may have initially been constrained by low oxygen levels. During the Ediacaran, oxygen began to increase, and when it reached critical levels in the Cambrian, animals were finally free to get big and interesting.

Like the M. C. Escher lithograph of hands drawing themselves, the main driver of evolutionary diversity might have been diversity itself. As soon as new animals evolved, they altered and built on their surroundings and interacted with each other, creating a feedback loop of new niches for even newer animal forms to adapt into.[8]

It all happened on the seafloor. Dry land, after all, was and would remain empty for a long time — no trees, no dinosaurs, not even any insects, just smears of algae and microbes. Even the bulk of the ocean was fairly empty; life within the water itself consisted of more algae, a few jellies, and a few little swimming worms. But where water met earth at the bottom of the sea, that's where the animal engineers went to work. Sponges began it, by filtering microscopic food from the water and transferring it to the ground, enriching the seafloor. Other animals colonized this newly hospitable space. Blessed with hydrostatic skeletons, they burrowed in. Their burrows brought food and water deeper underground, which allowed new kinds of microbes to grow — and become food for still newer animal forms. And as animals continued to explore new food sources, it was inevitable that they turned on each other.

Predation! It's one of the most fundamental biological interactions we know, from the spider catching a fly to the lion stalking an antelope. Yet there is almost no evidence of predation among those early, simple, strange animals, the Ediacarans. The American paleontologist Mark McMenamin of Mount Holyoke College has referred to it as the "Garden of Ediacara" in reference to the peaceful Garden of Eden.[9]

The very first animal predation shows up among the very last of the Ediacarans.[10] Fossilized tubes of nested cones named *Cloudina* from just before the Cambrian are sometimes found with holes in them.[11] Although the victims are more obvious than the perpetrators, the most likely suspect is some kind of arrow worm. These small but ferocious predators still terrorize the sea's wee beasties today. Arrow worms and their ilk may even have driven much of the early Cambrian diversification, as animals tried every trick to avoid getting eaten.

Shells, the hallmark of mollusks, probably came into existence as armor. The first Cambrian fossils after *Treptichnus* are called the "small shellies," and many are early snails, proto-snails, and snail relations. Like all the mollusks to come after, they built their shells with calcium carbonate. Calcium is life's preferred material for hardening both external and internal skeletons — it shows up in bones as calcium phosphate. Calcification cropped up quickly in numerous Cambrian groups, including early relatives of starfish and corals as well as mollusks, and defense seems the most likely reason.

A less obvious but equally effective tactic for dealing with predators is to simply grow larger: if you're big enough, the scary little worms can't eat you. Another option for anyone without the advantage of armor or size is to get out of town: if you can't defend yourself in one place, go colonize a new habitat.

Unfortunately for prey but fortunately for the diversity of life on Earth (and for the scientists who study it), predators can adapt too. It wasn't very long at all before much larger predators made their debut — notably, the "weird shrimp" *Anomalocaris*. This aptly named creature really was weird, not just for a shrimp but for anything alive at the time. In a world of small animals that crawled and burrowed and sat on the seafloor, or perhaps floated passively in the water, *Anomalocaris* grew up to 3 feet (1 m) long and could actually swim.

Being large and hungry, *Anomalocaris* would certainly have stimulated the evolution of all kinds of antipredator defenses. Mollusk shells got stronger. Their bodies got bigger. And some of them—the very first cephalopods—escaped *Anomalocaris*'s tyranny of the seafloor by rising up into the water column.

A New Use for an Old Shell

Observing an animal grow from a single fertilized egg into a creature of many cells, from small to large, from simple to complex, it's almost impossible not to draw parallels with the whole course of evolution. At one time scientists thought they could learn how an animal evolved over geologic time simply by watching it develop over its lifetime. A human embryo *does* look rather like a fish, then a lizard . . . We know now that the connection between evolution and development is more complex, more nuanced. Modern humans, lizards, and fish all evolved from a common ancestor, not one from another. But observing the similarities between their embryos still helps us to understand how it happened.

Likewise, the embryos of modern nautiluses and modern squid look quite similar—much more so than the adults do. Both embryos grow attached to a large ball of yolk, like an embryonic chicken, on which they sit like flat little caps. At the very crown of the cap is the part of the mantle that will grow a shell; spread out below and around this proto-shell is the rest of the animal, including a ring of adorable arm buds. These embryos don't yet look much like either squid or nautiluses, but they do look like mollusks of *another* kind: an odd little group of not-quite-snails called Monoplacophora. The name is Greek for "single shell bearers,"[12] and the single shells these creatures bear are flat caps resembling the caps of cephalopod embryos.

Monoplacophorans have no common name (like "snail" or "clam") because they are so rare today. In fact, they were known only as fossils until 1952, when a deep-sea trawl brought up living representatives in "one of the zoological sensations of the twentieth century," according to one mollusk expert.[13] Lest you write that off as an enthusiast's dramatization, consider the thrill of discovering a "living fossil" (hey, we thought this went extinct 400 million years ago!) and, furthermore, note that the living monoplacophorans proved strikingly different from the snails with which their fossil ancestors had been erroneously lumped. Unique among mollusks, monoplacophorans turned out to have multiple repeating sets of shell-attachment muscles, kidneys, and gills. Such repetition can be an adaptation to deal with low oxygen levels and is also a prominent feature of the trilobites. Perhaps not coincidentally, monoplacophorans and trilobites dominated the early Cambrian fauna.

Given the great abundance of monoplacophorans at the time, it's no surprise that one of them should be the ancestor of cephalopods. Embryology suggests it, and the fossil record confirms it. One of the current best guesses as to what links the two groups is the fossil *Knightoconus* (probably not itself a cephalopod, though its descendants may have been), whose shell grew into a tall cone rather than a flat cap. Critically, this shell was roomy enough to be divided into chambers.

Here's how scientists think that happened, in three simple steps. First: some monoplacophorans began secreting a liquid into their shells that was less salty than seawater. Like heating the air in a balloon to make it lighter than the surrounding air, freshening the water in a shell makes it lighter than the surrounding water. This may have made heavy shells easier to carry as their inhabitants continued to crawl on the seafloor. Second: some descendants of these first shell lighteners began to alternate the secretion of liquid with the secretion of more shell. Such periodic shell secretion would seal off chambers and prevent the fluid from leaking out. Third: descendants of the descendants used a thin tube of flesh, stretching back through every chamber in the shell, to extract liquid and replace it with gas. The extra buoyancy lifted both shell and animal up into the water.

You see, a shell's biggest drawback is its weight, which limits both size and mobility. Snails are not exactly renowned for their speed, and the largest non-cephalopod mollusk, the giant clam, lives a stationary life. A buoyant shell was *the* key innovation of cephalopods. It constituted a total transformation of the mollusk body plan, turning crawlers and oozers into floaters and swimmers. It was the first of many brilliant cephalopod innovations, before jet propulsion and instant camouflage and fancy eyes, yet it's all but forgotten today. We need paleontologists to provide the proper perspective. Björn Kröger of the Berlin Museum of Natural History calls the buoyant shell "an evolutionary step with a significance comparable to the development of wings in the insects."[14]

If that doesn't seem very significant, then take a moment to consider the world without bees or beetles, houseflies or butterflies, mosquitoes or midges, crickets or cicadas.

Like flight, buoyancy is a complicated trick—cephalopods couldn't just drive to the flower shop and hook up to the helium tank. For one thing, flowers wouldn't evolve for another 300 million years. The first necessary step was the development of a chambered shell, as we saw in *Knightoconus*. The walls separating the chambers are called *septa*, and they allow a cephalopod to seal off the gas-containing part of its shell from the living-in part, so the gas won't escape out the front door. The gas-containing part is called a *phragmocone*, from the ancient Greek *phragmos* for "fence," since it's fenced off. And in a surprising departure from science's infatuation with dead languages, the living-in part is simply called a *living chamber*. As a shelled cephalopod grows, it builds a new, bigger living chamber for itself and seals the old one. Thus, the phragmocone is really a series of dozens of chambers and septa.

When I first learned about the buoyancy of nautiluses, I assumed that they actively pumped gas into their shells, like a balloonist filling up for a flight. Had I lived in the nineteenth century, this notion would have been in line with all the available scientific

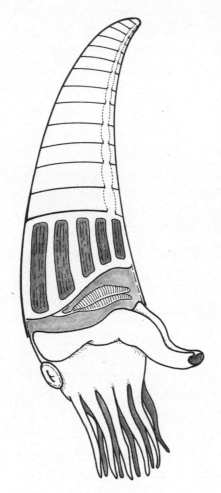

FIGURES 2.1A & 2.1B
It is generally accepted that
Plectronoceras cambria was the
first fossil cephalopod, with
its shell divided into chambers
and a narrow tube available
to adjust fluid content.
Opposite, two fossils with a
¼-inch (3 mm) scale bar; *left*,
an artist's reconstruction of the
living animal.
Fossil: Jakob Vinther;
reconstruction: B. T. Roach.

speculation. However, decades of careful observation in the twentieth century revealed that the nautilus concerns itself only with
pumping water out of its shell. Where the water used to be, gas
slowly seeps in, but this is simply an accident that illustrates the
difficulty of maintaining a vacuum.

Although most of the animal's body is in the living chamber, it
maintains a slight presence in the phragmocone to control the gas-
to-fluid ratio. A small tube of flesh called a *siphuncle* (based on the

Latin word for, unsurprisingly, "small tube") runs through every sealed-off chamber.[15] Remarkably, the animal can control the saltiness of the blood in its siphuncle, thereby taking advantage of water's tendency to diffuse across a membrane toward higher salt concentration, a tendency named *osmosis*. An extra-salty siphuncle naturally absorbs water from the phragmocone chambers, and the empty space left behind is filled with gases diffusing out from the blood.

This ability to continuously add new buoyancy chambers to their phragmocones allowed ancient cephalopods to reach truly stupendous sizes. Sure, shells got heavier as they got bigger, but cephalopods could always balance the weight with more gas. Evolution's ability to grow gigantic cephalopods may have been seeded by the first chambered shell, which was itself pretty minuscule.

The first of these potential floaters that we've found preserved in fossil form, complete with fluid-extracting tube, was less than

¾ of an inch (2 cm). The full name of this wee creature is *Plectronoceras cambria*. *Plectron* describes the shape of the shell, which is similar to a guitar pick, and *ceras* is a common "middle name" for all ancient shelled cephalopods. It means "horn," because shells and horns grow in a similar variety of shapes, straight or curved or coiled depending on the species. (The name *ceras* is so common that paleontologists have taken to abbreviating it with a *c*—for example, *Plectronoc.*) And the species name, *cambria*, of course, refers to the time when the animal evolved.

It seemed clear that all further cephalopods evolved from either the Cambrian guitar-pick horn or something very like it. That is, until someone raised the possibility that the critical invention of cephalopods was not buoyancy, but jet propulsion.

The Snail with a Jet Engine

Cephalopods were some of Earth's first swimmers, and they developed a rather unusual technique—the same mechanism that humans use to fly planes and shoot rockets. Jet propulsion is extremely rare in nature, and among those animals that use it, squid are the fastest by far. It's fairly straightforward: create a high internal pressure to expel a large mass through a narrow opening at great speed. Squid accomplish this by drawing water into the mantle through wide openings around the head, then sealing these openings and forcing the water out through the much narrower siphon.

Fossil shells show indentations that accommodated siphons, so we know this technique has been around for as long as cephalopods. The first siphon was probably just a folded mollusk foot, rolled into a tube as you might casually form a trumpet with your hands. Tooting your hand trumpet doesn't normally provide any significant propulsive power—unless you're in an environment with minimal gravity and friction, like outer space, where breath-

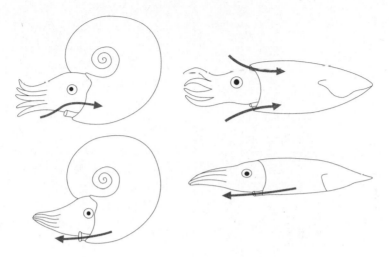

FIGURE 2.2 The difference in the jet propulsion of
an externally shelled and an internally shelled cephalopod.
C. A. Clark

ing out would be sufficient fuel to push your body in the opposite
direction.

Cephalopod jet propulsion, too, begins with a breath. Pumping water over the gills becomes fueling up; expelling respiratory
waste becomes propelling the body through water. A curious effect of this process is that cephalopods cannot breathe without
also moving.

None of the early shelled cephalopods would have been fast
swimmers, but at first it didn't matter, because there were so few
other swimmers. Cephalopod jet propulsion was good enough for
getting around in a world of trilobites, worms, and not much else.[16]

But in 2010, a couple of paleontologists made a paradigm-
shifting suggestion. What if the first cephalopods had no shells at
all and looked instead like modern squid?

This idea was shaped around the Cambrian fossil *Nectocaris*,
which has been known informally since the early 1900s from a

FIGURES 2.3A & 2.3B
Nectocaris, the "swimming
shrimp" — or swimming
squid? Argued by some to
predate even *Plectronoceras* in
the cephalopod family tree,
further evaluation has set it
apart as a curious example of
convergence. *Left*, fossil with
1-cm (10 mm) scale bar; *below*,
an artist's reconstruction.
*Fossil: Martin R. Smith;
reconstruction: Marianne Collins*

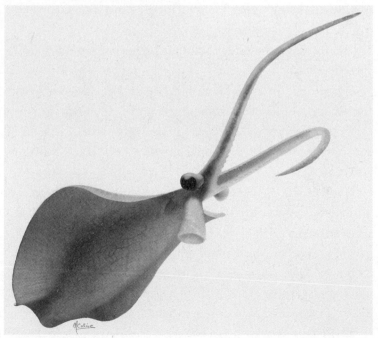

single specimen. Larger than *Plectronoceras* and predating it in the fossil record—*Nectocaris* is found in the early Cambrian, *Plectronoceras* in the late—*Nectocaris* was still small enough to fit on the palm of your hand. It had an oblong body ringed with fins and a vague resemblance to shrimp. It was named accordingly "swimming shrimp" (in Greek) and its exact affiliation left as a question mark.

More extensive fossil collection in the 1980s and '90s yielded ninety-one new *Nectocaris* fossils, which proceeded to sit in the Royal Ontario Museum, waiting for scientific attention. It came in 2008 when a graduate student named Martin R. Smith arrived at the University of Toronto interested in the early evolution of eyes. Smith recalls that his adviser, Jean-Bernard Caron, pointed him at the pile of *Nectocaris* and said, "'Here's a fossil that's not really been described; it's got eyes—why don't you take a look at it?' I said, 'Right, sounds good, looks exciting'—and it ended up being a much bigger story than we expected."[17]

Far better preserved than the original, the new *Nectocaris* fossils clearly possessed two tentacles, two eyes, and, crucially, a tubular structure that Smith and Caron identified as a siphon. Using these features, the two scientists argued that *Nectocaris* was an early cephalopod and specifically, given its lack of an external shell and similarity to squid, an early coleoid.[18]

With this Smith accomplished a rare feat for a graduate student: publication in the high-profile journal *Nature*. It caused, if not exactly an uproar, certainly some consternation among the paleontologists of the world. After all, the first definite coleoid fossils date to 200 million years after *Nectocaris*, and Smith concedes that the idea of a coleoid way back in the Cambrian is "problematic." What's more, the anatomy of *Nectocaris*, which looked so convincingly cephalopodan to Smith, has drawn abundant skepticism.

Some scientists pointed out that the supposed "siphon" would be useless for swimming, as it expanded out like a trumpet rather than constricting to produce a jet stream. They suspected that it was instead some kind of slurping proboscis.[19] Others focused their critique on the creature's body itself. Is the space seen within the body of *Nectocaris* fossils the mantle cavity of a cephalopod, open to the surrounding sea? Or is it simply a gut, busily digesting food?[20]

Smith, now at the University of Durham, recalls wryly, "There was a bit of backlash when the paper was first published." He agrees that "the most important character is this axial cavity" but, alluding to other scientists' suggestions that it was a gut, says, "I don't think those hold water." (The pun was either unconscious or the delivery so deadpan I couldn't tell.)

"If you buy that it really is an axial cavity that was used in jet propulsion, that's something we only see in cephalopods," Smith points out. "If you want to exclude *Nectocaris* from the cephalopods, you've got to say this is a convergent feature. It's always possible, you know—bats have wings and birds have wings. Convergence is a theme in evolution."

Convergence happens when two unrelated groups of organisms arrive at analogies in anatomy (or biochemistry, physiology, or behavior). Flight is a quintessential example. Another is the convergent evolution of wolf types and lion types in both placental and marsupial mammals. Similarly within the soft-bodied shell bearers, slug types and limpet types have evolved numerous times in distinct lineages. And then there's the panoply of convergences between cephalopods and fish: eyes, body shape, muscle fibers, nerve fibers, and more.

Nectocaris seems to have formed an early precedent for such convergence, long before fish came along.

In 2011, motived in part by Smith's provocative paper, a group of three young European scientists published a review of the entire evolutionary history of cephalopods. Björn Kröger and Dirk Fuchs were working in Berlin, Kröger at the Museum für Naturkunde, and Fuchs at Freie Universität; their collaborator Jakob Vinther was a graduate student at Yale University. We'll be hearing more from all three over the course of this book, but for now let's consider that their review paper, "Cephalopod Origin and Evolution," has been cited and referenced so often that it seems to be forming a new foundation for the field.[21] The news that true nautiloids are not as ancient as we thought is thanks to Kröger, Vinther, and Fuchs. So is the latest picture of how cephalopods got from shelly ancestors to shell-less descendants.

In their review, the authors devote a full page to the creature they called "*Nectocaris*: a lost child of the Cambrian." Working through each trait, they conclude that *Nectocaris* is probably not even a mollusk. But they agree that some features are reminiscent of squid and suggest that this ancient creature's lifestyle was "remarkably similar to [that of] cephalopods."

Nectocaris's convergence with the successful shape of a modern squid invites questions about the nature of the Cambrian ecosystem. Why did *Nectocaris* and its relatives disappear before they hardly got started? Were they simply unlucky, the losers in chance events that might as easily have left their descendants alive today to duke it out with squid—and fish? Or were they, perhaps, some of the first victims of the first cephalopods?

Superpredator

The changes wrought by cephalopods rising from the seafloor were certainly profound. The American paleontologist and nautilus biologist Peter Ward has made the case that all the creatures grubbing around in the mud, especially trilobites, were literally blindsided by this development. Their eyes couldn't even look above them, because they had no reason to. In the wake of the cephalopods' arrival, he argues, trilobites evolved upward-looking eyes and upward-pointing spines to defend themselves.[22]

"Like owls snatching mice, death came from above for the bottom living invertebrates of the Cambrian period, and it came rapidly and without warning," wrote Monks and Palmer in *Ammonites*.[23] Monks has also called these cephalopods "the great white sharks of their day."[24]

The idea of early cephalopods as superpredators is a pleasing story for us sucker lovers to buy into. But cephalopods may not have been truly the first in size or position in the water—after all, *Anomalocaris* could swim and gobbled up trilobites right and left. What cephalopods did, and did quickly, was dominate this niche. *Anomalocaris* and cousins faded from the scene after the Cambrian, and the Ordovician period that followed saw a great flowering of cephalopod diversity.[25]

Early cephalopods had long straight shells, mostly between 12 inches (30 cm) and 6½ feet (2 m) in length. But one species, appropriately named *Endoceras giganteum*, grew up to about 12 feet (3.5 m) long—taller than a basketball hoop and far bigger than any *Anomalocaris*, larger indeed than any living creature the world had yet seen.[26] Its shell became so buoyant that *Endoceras* had to backfill the first chambers with heavy minerals, presumably deposited by the siphuncle. This offset the weight of the body at the other end and allowed it to swim horizontally like its smaller cousins, instead of bobbing around like an awkward exclamation point.

These early cephalopods would have been, shall we say, *stately* swimmers. They were so very stately that, until evidence to the contrary surfaces, the German paleontologist Dieter Korn is inclined to think of them as plankton rather than active swimmers. The word "plankton" usually evokes microscopic representatives like single-celled plants, but it also refers to the great jellies and salps that can grow longer than a person—and possibly some of the largest fossil cephalopods.

Korn works at the same Museum für Naturkunde as Björn Kröger, was in fact Kröger's postdoctoral adviser, and has been called the grand old man of cephalopod paleontology. He reminds us, "There was no need for [early cephalopods] to swim quickly, because there were no large fish. I imagine it was like paradise for them, all these little critters. There was not a lot of competition. Lovely for such an animal."[27]

Given the other fossils found most abundantly from the same time period, it's perhaps inevitable that generations of artists have depicted cephalopods hunting trilobites. But Korn objects. "First one has to ask, is it really fun to eat a trilobite? There aren't many nutrients in a trilobite; it's mostly a calcite carapace."

Another serious problem with this scenario is that beaks, which are used by all modern cephalopods to support their predatory lifestyle, are conspicuously absent from this part of the fossil record. "How are you going to crack a trilobite shell if you don't have a beak?" asks the University of Zurich's Christian Klug (who was one of Korn's graduate students).[28] No fossil cephalopod beaks have been found from the first 100 million years of these animals' existence, although radulas have been described from a few of the earliest cephalopods.[29] Says Klug, "If there's a radula you should have the beaks as well, but you don't see the faintest trace."

Of course, Klug acknowledges, "that's not a proof that they weren't there." The scientists who described many of the early cephalopods simply weren't looking for beaks, and it's hard to find something you're not looking for. Even among later cephalopods like ammonoids and coleoids, beaks are rarely found. And the strong similarity between the beaks of living nautiluses and living coleoids indicates they probably evolved from a common ancestral beak. Perhaps we've simply not been lucky enough yet to find any such ancestors.

With or without beaks, the first cephalopods must have eaten *something*, or their evolution would have ended then and there. If they weren't crunching through trilobite armor, then what? Klug points out, "If you have that huge size, it makes you think

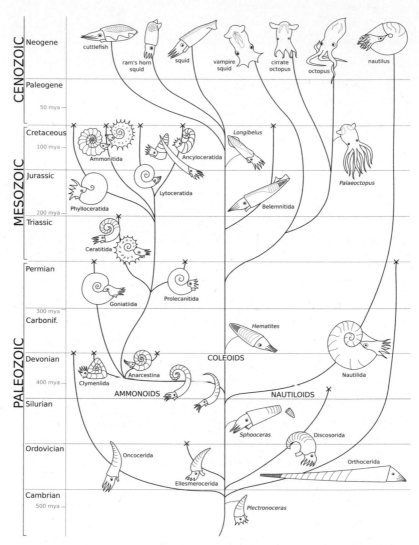

FIGURE 2.4 The evolutionary history of cephalopods is here mapped onto the earth's geologic timescale. Species of squid, cuttlefish, and all the rest that we see in the ocean today are merely the latest descendants of their storied lineages — they themselves were not around in the Triassic! Internal anatomy is drawn in coleoids to illustrate the steps of shell reduction, and it is omitted in externally shelled cephalopods to show shell ornament. Dotted lines indicate genera whose stories are told in the text; all other lines represent substantial taxonomic groupings.

Danna Staaf and C. A. Clark

about filter feeding." After all, the largest animals alive today are the filter-feeding baleen whales. "Maybe back then they were just taking in small bits and pieces, both from the ground and from the water column."

A distinct but related possibility is the scavenging lifestyle, which involves the consumption of *large* bits and pieces. If, for whatever reason, early cephalopods weren't equipped to hunt and kill their prey, they could simply have waited for time, disease, or accidents to do the work, then bobbed along afterward to pick up the carcasses. Says Korn, "I would say this is most likely, that they were scavengers."

Still, we needn't entirely abandon our notion of early cephalopods as predators. Monks, for one, favors the idea of predation in slow motion. Let's take a moment to imagine a large straight-shelled Ordovician cephalopod drifting along, several feet above the sandy bottom. It senses—perhaps with vision, perhaps with smell—that a snail is crawling below at an equally leisurely pace. With a few gentle jets the cephalopod pushes itself within arm's reach and collects the prey.

If filmed for a nature documentary, this "chase" scene would have to be played back triple speed to keep the audience's interest, but it would nevertheless constitute a real chase.

We must remember that the world is large and heterogeneous. Everything didn't evolve everywhere. In some places the early cephalopods may have floated passively, while in others they developed more active swimming techniques. Evolution is not a single thread—it's the interweaving of many, many threads, some cut short, others so quickly changed you can hardly follow them through the cloth.

Whether they were superpredators, superscavengers, or superplankton, early cephalopods were still top dog. Until fish came along.

Early Experiments

While cephalopods large and small spread throughout the Ordovician oceans, hunting, filtering, or scavenging (or perhaps all
three), a rather tangential and unassuming lineage called *vertebrates* produced the first proper fish. They had fins and tails and
gills; they even had skulls — but no jaws. They slurped up whatever food didn't need biting or chewing and were therefore almost
certainly no threat to cephalopods. Then, in the subsequent Silurian period, these odd bony creatures evolved something truly
dangerous.

Jaws were just as effective as beaks for catching and eating prey.
Jaws could even puncture or crush a cephalopod's shell. Silurian
cephalopods now had competition for their prey, and they may
well have had their first experiences of being turned into prey
themselves. The new fish on the block presented cephalopods
with evolution's grim mandate: adapt or die.

All these early jawed fish wore armor that looks rather clunky
to us today — it certainly wasn't the most hydrodynamic style.
They were no sharks and they were no marlins. Still, they were
nothing nearly as clunky as cephalopods, who had to truck around
massive molluscan shells. Fish armor was at least articulated, able
to bend and move. Cephalopod shells were less armor and more
mobile bomb shelter.

To cope with this disadvantage, some Silurian cephalopods
took a drastic approach, periodically breaking off the awkward
tips of their long straight shells. Such a truncated shell would
be easier to maneuver while the cephalopod was swimming and
less likely to be damaged if the animal was caught up in waves or
bumped into rocks. Although the advantages are clear, deliber-
ately breaking one's own shell is such a curious strategy that the
earliest known truncating cephalopod is named for it: *Sphooceras
truncatum*.[30]

The Bohemian paleontologist Joachim Barrande, who first described *Sphooceras* fossils in 1860, estimated that truncation could happen dozens of times during the animal's life.[31] *Sphooceras* began by filling up three to five chambers at the end of its shell with hard mineral deposits, like the counterweights we saw in their enormous cousins. Similar deposits were used to thicken the wall, or septum, between the part of the shell that *Sphooceras* would keep and those chambers destined for removal (delightfully named "deciduous," like the leaves that fall from a maple tree). Finally, the siphuncle was plugged and the deciduous chambers were disconnected. Somehow. Mysteriously. We'll come back to that in a moment, after mentioning another mystery:

The septum that once connected to the deciduous shell grew a smooth cap after being exposed to the sea. Such a cap would have to be created by the animal's living body—which was all the way at the other end of the shell. Barrande came up with a pretty wacky idea to explain this: he suggested that *Sphooceras* could stretch two special tentacles back along the length of its truncated shell to secrete the cap.

It wasn't until the 1980s that several researchers took issue not only with the tentacles-secreting-shell-cap idea but with the entire concept of shell truncation. "One cannot suggest any reasonable mechanism," objected Polish paleontologist Jerzy Dzik.[32] The animal might be able to fill chambers and thicken a septum, but then what? What magic would cause separation at exactly the predetermined point? With the animal living at the opposite end of its shell from the truncation site, Dzik could not imagine how it could break its own shell, and any "extrinsic factor" that could break the shell would also kill the animal.

Then in 2012, the Czech scientists Vojtěch Turek and Štěpán Manda and a couple hundred new *Sphooceras* fossils came to the rescue of the truncation hypothesis, with a whole new perspective on a very old animal.[33]

FIGURE 2.5 *Sphooceras truncatum*, possibly the first
cephalopod to internalize its shell.
Franz Anthony

Consider a cowrie, unique among seashells for its naturally
polished exterior. Other snails keep their whole bodies, including
the shell-secreting mantle, inside their shells, so the outsides grow
gradually rough and weathered. A cowrie maintains its smooth
texture and beautiful patterns by periodically reaching its mantle
out to engulf the entire shell and secrete new material.

Even after millions of years buried in rock, the shell of *Sphooceras*, and most notably its cap, show cowrie-like polish and pattern. Turek and Manda concluded that the animal's mantle must
have stretched all around its own shell, leaving telltale clues for
paleontologists to decipher. Further evidence for this theory came
from the other end of the shell, the opening near the head, where
the shell seems so thin that it would have broken if not protected
underneath a layer of skin.

With this new view of *Sphooceras*'s mantle covering most or all
of its shell, truncation becomes easier to understand. The mantle
could just start dissolving the shell at the site of intended break-
age. Like perforating a piece of paper, this would weaken the deciduous material enough for it to simply fall off under the routine
stress of swimming. Of course, the mantle would have to be retracted in order for truncation to occur — otherwise the deciduous shell would be retained within the body, completely defeating
the purpose.

That the shell would have been exposed during at least some portion of the animal's life is also indicated by a color pattern on the best-preserved specimens. Turek and Manda describe the pattern as similar to that of a modern nautilus, with thick stripes on the top side of the shell.

The tiger stripes of modern nautiluses seem flamboyant out of context in a gift shop, but when viewed underwater they form a type of camouflage called *countershading*. Dark stripes lie thick and close on top of the shell, so as you look down from above, the nautilus blends in with the dark ocean depths. On the bottom of the shell the stripes are thin to nonexistent, so the animal blends in with the bright sea surface to anyone looking up from below. A comparable pattern on *Sphooceras* suggests that its shell would have been at times exposed and therefore worth hiding from predators (or prey).

Although Dzik still isn't convinced that *Sphooceras* necessarily enwrapped its shell with the mantle and engaged in truncation, he calls Turek and Manda's study "one of the best executed works on Palaeozoic cephalopods."[34] By discovering, documenting, and analyzing so many new *Sphooceras* fossils, these two scientists have irrevocably advanced human understanding of our planet's past. Observations can be used forever, though interpretations might be refined or overturned.

If Turek and Manda's interpretations hold up, then *Sphooceras* may have been the first cephalopod to experiment with internalizing its shell. Many millions of years later, a few renegade ammonoids might give it another try, but it wouldn't be until the first coleoids jetted onto the scene that internal shells would really take off. These very first coleoids might have even briefly revived the truncation tactic, long after its pioneer had gone. Poor *Sphooceras*, adapted to sharing the Silurian seas with early jawed fish, was not prepared to survive the Revolution.

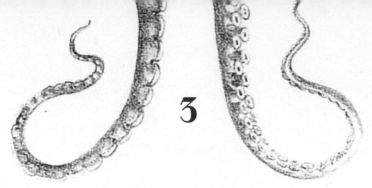

3

A Swimming
Revolution

Despite evidence of shell shedding in *Sphooceras*, truncation remained a little-used technique for dealing with the unwieldy cephalopod shell. The mainstream approach became coiling the shell, as one might wrap long hair into a bun. The first coiled forms showed up within a few million years of the first cephalopods, swimming alongside straight-shelled forms. As time went on, coiling continued to evolve independently in various groups. One was the nautilids—that's not a typo, but a specific subset of nautiloids that would give rise to today's pearly nautilus.

The other animals with notable coils were the ammonoids.

Ammonoids are a fascinating group. Their abundance and diversity dominate the fossil record of cephalopods, yet they left not a single soft-body trace or living descendant to guide our imagination. From their shells alone we must piece together their origin in the fishy Devonian seas, their evolution of a prolific spawning habit and a hasty race to maturity, and their near-total demise at the close of the Paleozoic era.

Jaws

The Devonian period of 400 million years ago has often been called the Age of Fish. Such a title implies that everything non-fish was toast, and the implication is not entirely inaccurate. But the Devonian also hosted a tremendous diversification of cephalo-pods, nautiloids as well as ammonoids. Both fish and cephalopods continued to get better and better at swimming—and better and better at eating.

The first jawed fish had evolved back in the Silurian, but it wasn't until the Devonian that they really went wild, producing updated versions of the awkwardly armored *placoderms* as well as a number of sharks and bony fish. Then, it seems, natural selection began to exert a strong pressure toward hydrodynamic shapes, fa-voring fish that could cut through the water with speed and agil-ity. Over the course of a few million years, their bodies shifted from strange flat pillowy things to what we think of as a proper fish shape, streamlined like a tuna.

A similar transition occurred in the cephalopods. Straight-shelled or only loosely coiled cephalopods became less diverse and less abundant by the end of the Devonian, while those that built their coils tighter and tighter flourished. Coiling confers the same advantages in speed and maneuverability as truncation, with added protection against being eaten. A coiled shell is harder to grab, harder to hold, harder to break. Cephalopods needed that defense.

The ocean ecosystem had been overhauled. A water column that once housed mostly small drifters and large but ponderously slow cephalopods was now a highly competitive racecourse, with the losers facing death by jaws.

To understand what happened, it helps to learn a bit of marine biology jargon. Animals can be classified according to home and habit as bottom-dwelling *benthos*, drifting *plankton*, or actively swimming *nekton* (from the same Greek root as our friend *Nectocaris*, the swimming shrimp). Most early animals back in the Cambrian had been benthos. Cephalopods were some of the first to move into plankton and nekton, radiating into new niches throughout the next two periods, the Ordovician and Silurian. Now the Devonian was seeing an evolutionary shift in all animals, regardless of taxonomic affiliation, out of both benthos and plankton and into the nekton.

Thus, in 2010, the Swiss paleontologist Christian Klug coined a new term: instead of the Age of Fish, he proposed the Devonian Nekton Revolution.[1] "The pressure on the benthic organism was rising," says Klug. The Devonian saw "the evolution of spiny trilobites, spiny echinoderms [sea urchins] — evolution of defense mechanisms happened across all groups."[2] Survival had been hard enough when it was just predatory cephalopods drifting along above the bottom, snatching prey at their leisure. But now these hungry fish with their cracking, crushing jaws had crashed the party, and all the prey species had to adapt.

Not coincidentally, the Devonian also gives us the first fossil evidence of cephalopod beaks.[3] "It could be that jaws evolved in cephalopods as a reaction to the evolution of jaws in fish," speculates Klug.[4] That doesn't mean cephalopods and fish were sitting side by side comparing mandibles, and cephalopods felt inadequate so they went home and built beaks. No. It means that various prey species were evolving better armor and stronger defenses in response to fish jaws, and any cephalopods without comparably powerful chompers were going hungry. Those cephalopods that happened to grow harder beaks were able to catch and eat more food. Well-fed animals tend to reproduce more successfully than hungry ones do, so hard-beaked cephalopods left more descendants, eventually leading to a sea change in cephalopod mouths.

Vinther points out that the relationship could also have gone in the other direction. "Who knows who started the whole thing?"

he says. "Whether it was a jawed vertebrate that made everybody go, 'Shit, okay, we have to adapt,' or whether it was the cephalopods that evolved jaws and the fish realized, 'Okay, we need to keep up.'"[5] The fossil record hasn't yet been dug up and brushed off in enough detail for us to be sure.

However it started, the Devonian Nekton Revolution marked the beginning of a relationship that would last all the way through to the modern day. Fish and cephalopods, cephalopods and fish, both diversifying to fill the oceans, competing with each other and eating each other. "That seems to be the theme in cephalopod evolution," says Vinther, "that they are constantly coevolving with fish."

The importance of fish in the evolutionary trajectory of cephalopods is indisputable, but there's some scientific hemming and hawing over whether interactions between the two groups are primarily competitive or predatory. Do cephalopods and fish jostle each other for the same prey—or are cephalopods just trying to escape fish jaws? Dieter Korn, for one, contends that the two groups never really met on equal footing. Cephalopods were hampered by their shells, among other things, and "an ammonite was probably not able to do any sort of hunting of other animals, except maybe for very small ones. As long as there were fishes of any sort, they were top predators, not the cephalopods."[6]

This viewpoint offers some validity to the old "Age of Fish" moniker, at least if you're inclined to name a time period after its top predators. But there's also a case for naming it the "Age of Ammonoids," after the creatures that may well have taken up a central position in the marine food web. This could have been when cephalopods began to serve as ecological keystones, devouring the ocean's smaller edibles and providing ample food for its largest denizens. Just as squid offer a one-prey-fits-all solution to modern marine predators, ammonoids might have done the same in ancient seas.

Such a role would have been facilitated by a significant evolutionary advantage ammonoids developed over their forebears: breeding like rabbits.

Sex and Babies

This is one arena in which cephalopods, both ancient and modern, are actually less alien than many animals — even other mollusks. Slugs, for instance, are hermaphroditic, and in the course of impregnating each other their penises sometimes get tangled, so they chew them off. Nothing in the rest of this chapter will make you nearly that uncomfortable.

All living cephalopods have separate male and female sexes, so we assume that extinct cephalopods did too. But it's awfully difficult to identify the sex of fossils — it can be tricky even with live animals. A male nautilus is known by his *spadix*, which frankly should just be called a penis since it's erectile and serves to transfer sperm to the female. Male coleoids are a little more reserved; they keep their penis equivalent inside their mantles and use a modified arm called a *hectocotylus* for sperm delivery. Both spadix and hectocotylus are hard to recognize unless you know exactly what you're looking for, and in most species the only way to identify a female is to be absolutely positive that she doesn't have them.

In one living coleoid species, however, sex is blindingly obvious. Females of the octopus known as an argonaut are *five* times larger than males. (A killer whale is about five times larger than an average adult human, which in turn is about five times larger than an opossum.)

This enormous size differential caught the attention of paleontologists who had noticed that many ammonoid species also came in two distinct sizes, which they had dubbed microconch (little shell) and macroconch (big shell). Both were clearly mature, as they had completed the juvenile part of the shell and constructed the final adult living chamber. After an initial flurry of debate, most researchers agreed to model ammonoid sex on modern argonauts, and began to call macroconchs females and microconchs males. Microconchs also often have elongate protrusions called *lappets*, which could have provided some, ahem, anatomical support (the hectocotylus of certain modern coleoids can be quite large, and it's been postulated that the microconch hectocotylus was similarly oversized).[7] Or the lappets might have been more like a peacock's tail, a decoration to attract mates.

Some fossil nautiloids also come in macroconch and microconch flavors, though it's more difficult to be certain that both are adults. Identifying adult ammonoids is usually straightforward, as many species developed elaborate modifications of the final living chamber — like those lappets. Signs of maturity in nautiloid shells consist of more subtle changes in relative thickness of the septa and size of the chambers. Still, in a few fossil nautiloids, scientists have gone ahead and labeled the nautiloid macroconch "female" and the microconch "male."

However, the shells of modern nautiluses show the opposite pattern — males are somewhat larger than females, with a wider aperture to accommodate the spadix. Like the nautiloid shift from ten arms to many tens of arms, this pattern could certainly have evolved from a different ancestral condition. If we're going to make that argument, though, we have to wonder when nautiloids switched from females to males as the larger sex, and why.

In modern species that have larger females, we usually assume the size difference has to do with making or brooding a lot of eggs. Female argonauts take it up a notch and actually secrete a shell-like brood chamber from their arms, using it to cradle numerous batches of eggs over their lifetime. Meanwhile, each tiny male argonaut gets to mate only once. His hectocotylus is disposable — after being loaded with sperm and inserted into the female, it breaks off. Although every other male coleoid we know of has a reusable hectocotylus, they're all named after the poor argonauts, because the first scientist who found a detached arm inside a female argonaut described it as a parasitic worm, genus *Hectocotylus*. Only later did biologists realize that the "worm" had originally belonged to the male argonaut and that other male cephalopods bore similarly modified arms.

By contrast, when males are the bigger sex, we often guess that the purpose is competition. Certainly many species of squid and cuttlefish have large males that battle for female attention on the mating grounds. They display outrageous skin patterns as they push, shove, and bite each other. Females do appear impressed; at least, they mate with the winning males and consent to be guarded by them. Even in these species, though, there are some small males, who exhibit a totally different mating strategy. While the big males strut their stuff, these small males quietly sidle up to the females, sometimes disguising themselves with female color patterns. This doesn't put off the real females, which readily mate with these aptly named "sneaker males." By their very nature, such obfuscating tactics are virtually impossible to glean from the fossil record. The imagination can run wild, but how would we ever know if ancient male cephalopods came in two sizes?[8]

Regardless of size differential or mating tactic, females of all the modern cephalopod species that have been studied collect and store sperm from multiple males before laying their eggs. This system is both fascinating and frustrating for biologists. It offers abundant opportunities for the sperm to compete with each other and for females to choose sperm from their favorite males, but the actual outcome is a mystery. First sperm wins? Last sperm wins? Fastest sperm or sperm from the sexiest male? Or maybe all the sperm share the available eggs among themselves? These are hard questions to answer with live cephalopods, let alone fossil ones. Sex cells don't exactly fossilize well.

Yet paleontologists are lucky enough to have discovered a few fossil cephalopod eggs. (No fossil sperm yet.) The diversity of these discoveries suggests that ancient cephalopods might have followed as many different spawning strategies as their descendants do today. Some ammonoid eggs appear to have been packaged in capsules, like those of modern octopuses, cuttlefish, and some species of squid.[9] Both modern and ancient egg capsules are found attached to relatively stable objects, like rocks or algae or empty shells, usually on the seafloor. Other ammonoid eggs have fossilized abundantly in places where the seafloor would have been deadly due to lack of oxygen. Thus, paleontologists surmise that these eggs originally floated far above this inhospitable environment, in the well-oxygenated water closer to the surface. Such free-floating gelatinous masses are commonly laid by oceanic squid today.[10]

A few collections of ammonoid eggs have even been found inside macroconchs, which is tempting to interpret as brooding behavior and further support for calling macroconchs females.[11] Though it's a rare habit among modern cephalopods, a couple of deep-sea octopus species do hold developing eggs inside their mantles until the eggs hatch.

Lucky finds of fossil eggs are too rare to illuminate overall evolutionary trends in the sex lives of ancient cephalopods. For that, scientists rely on an extremely handy feature of adult mollusk shells: each shell constitutes a record of the animal's entire life, including the size of the egg it hatched from. Within an ammonoid fossil, at the center of its coiled shell, lies the tiny *ammonitella* that the animal grew while it was still an embryo. (I apologize for burdening you with yet another obscure term, but "ammonitella" is so delightful.)

As Yacobucci explains, "The ammonitella forms all in one piece in the egg, and right after the animal hatches and pops out of the shell, there's a mark on the shell of 'here's where it hatched.' They do not have a larval stage, which is unusual for marine invertebrates.

What hatches out of the egg is this little miniature adult." She also notes, "The ammonitella often has a distinctive pattern of ornamentation which is different from that of juveniles or adults."[12] This trifling detail will turn out later to hold vital evolutionary clues. Evo-devo, after all, shows us how much evolutionary mileage an organism can get from patterns that change as it develops. For now, though, let's focus on the fact that measuring the ammonitellas of adult shells tells us how big they were as babies, and analyzing the rest of the shell can tell us how quickly they grew.

Tracking the growth of ammonoids in this way over the course of the Devonian gives us the Case of the Incredible Shrinking Eggs. Like many a good mystery, the story begins far from the scene of the action, with the growth on land of Earth's first forests.

The Incredible Shrinking Eggs

Around the beginning of the Devonian, land plants began to evolve from algal scum and creeping moss into real three-dimensional structures. They dug deep with their roots and spread wide with their leaves, and eventually grew tall enough to be called trees. Colonizing all the available dry land, these trees dropped bits and pieces of themselves into streams and rivers, which eventually washed everything out to sea.

The abundant influx of edible matter caused tiny drifting plankton to bloom in abundance. Not all plankton could make use of decomposing leaves, certainly, but as those that could proliferated, they became food for other kinds of plankton. This surge of energy from the bottom of the food web fueled evolutionary radiation throughout.

So far, so good. It's fairly straightforward to see how this sequence of events could have worked in synergy with the evolution of jaws to fuel Klug's Nekton Revolution. And then Klug's graduate student Kenneth De Baets took the line of inquiry in a new direction.[13] Sure, ammonoid shells as a whole got progressively more coiled over the Devonian. But so did the tiniest part of each ammonoid shell, the ammonitella. Furthermore, as ammonitellas coiled more tightly, they also got smaller. What was that about?

De Baets is yet another cephalopod sucker who started out as a dinosaur dude. Once his initial interest in dinosaurs brought him into the study of geology, he became more and more intrigued by invertebrates. Watching a documentary one day, De Baets heard the famous story, oft told and retold, of the octopus that crawled out of its own aquarium and into other nearby fish tanks to dine on its neighbors.[14] It got him wondering about the day-to-day lives of the octopus's far-distant ancestors. Among other questions, he wondered about their reproductive habits.

He knew that in modern octopuses, species with larger eggs lay fewer of them, while species with smaller eggs lay more of them. The difference in egg quantity can span several orders of magnitude, and when he tackled the study of ancient ammonitellas, De Baets found a similarly dramatic span: from estimates as low as 35 large eggs per female to estimates as high as 220,000 small eggs per female. Although variation between ammonoid species continued throughout the Devonian, smaller eggs and more prolific parents tended to occur later in time.

Since their Cambrian origins, most cephalopods had been laying relatively few large eggs, full of nutrient-rich yolk to sustain their developing offspring. It was necessary to do so, as not much baby food was available in the environment. Starting in the Ordovician, some of the early straight-shelled cephalopods might have experimented with laying smaller eggs containing smaller babies that would need to fend for themselves among the plankton. But it was the great plankton blooms of the Devonian that made it consistently worthwhile for cephalopod parents to send their babies out to eat rather than pack them expensive yolk lunches. Ammonoids laid smaller and smaller eggs, which hatched quickly into plankton-gobbling babies — and the smaller the eggs got, the more of them each female could lay.

Unfortunately, as a side effect of smaller eggs, the hatchlings became vulnerable to predators of a greater size range. The defensive tightening of ammonitella coils likely came about as a result of this vulnerability, just as increased predation pressure was selecting for tighter coils in adults.

You might wonder whether smaller babies would take longer to grow up — whether a reduction in egg size would lengthen the animal's life span. That doesn't seem to have been the case. Tracing a shell from its embryonic center to its final body chamber can help us figure out how quickly the animal matured, and the answer for most ammonoids is: very quickly. Some species might have matured in as little as a year, others in five, or ten at most.

Ammonoids, then, had converged on the same lifestyle as most modern coleoids, which are famous for living fast and dying young. Today's squid lay thousands, sometimes millions, of eggs and die immediately afterward, never meeting their own offspring. In most species, the babies grow up to spawn their own eggs and die in less than a year. Ammonoids and squid are hardly the only animals to adopt this strategy: most insects do it and, surprisingly, so did another exceedingly successful but now-extinct group: dinosaurs.

Dinosaur lives were a good deal longer than those of mosquitoes or squid, but their maturation was quick compared with that of other vertebrates — especially the earliest mammals. Each new generation offers natural selection a new chance to work, and with a more rapid generational turnover than mammals, dinosaurs simply evolved more quickly. They adapted, spread, and diversified, sidelining mammals for more than a hundred million years.[15]

The story of dinosaurs' early ascendance on land is mirrored by that of cephalopods in the sea, where ammonoids prospered for geological period after period — and rebounded quickly from the first of many serious setbacks, the end-Devonian extinction.

This crisis may have been created by the same influx of vegetative material that formed a critical spur to early ammonoid evolution. We see in today's oceans that large quantities of nutrients washed into the sea, such as fertilizer runoff from agricultural operations, can have a toxic effect on the ecosystem. As marine bacteria work to digest the excess nutrients, they absorb too much oxygen from the surrounding water, and animals that need to breathe find they can't.[16] A similar cascade of events is thought to have occurred at the end of the Devonian, when a widespread extinction caused the total collapse of massive reef systems alongside the disappearance of numerous ammonoid lineages.[17]

Thanks to their abundant offspring and short generations, it didn't take long for the surviving ammonoids to diversify yet again in size and shape throughout the subsequent periods (Carboniferous and Permian). It seems likely that their speedy evolution overshadowed that of the nautiloids, which remained minor, background players through the second half of the Paleozoic. Nautiloids continued to lay fairly large eggs like the Ordovician cephalopods of yore and matured to adulthood at a leisurely pace. Consequently, their evolution was unhurried and the diversity of their shells was limited, hemmed in on all sides by an astounding array of ammonoids.

It is really no surprise that ancient nautiloids grew slowly—modern nautiluses certainly do, constrained by the demands of shell construction. Come to think of it, ammonoids had to build shells too, so how could they have grown up so much faster than nautiluses?

The Ammonoid Cheat Code

Some of the most beautiful and dramatic features of ammonoid fossils are the "suture lines" that trace the meeting between septa and shell. The more looped and recurved the suture line, the more wrinkled and folded the septum, and there's no shortage of scientific speculation as to the purpose of all this complexity. The most popular theory relates the intricacy of the suture with the strength of the shell to resist pressure.

Any submerged container of gas runs the risk of implosion from water's incredible pressure. As a human, you probably don't worry about this much—humans are mostly water and water is mostly incompressible. We keep air in only a few places; our squishy lungs compress painlessly, and we can equalize the more rigid sinus spaces in our heads by adding air from our lungs. You feel this happening when you dive to the bottom of a swimming pool, where your ears will probably pop—if they don't, you can help them along by pinching your nose and blowing.

Shells can't be "popped" because cephalopods don't have another reservoir of gas to exchange with. So what's an ammonoid or nautiloid to do? One option is to stay in shallow water near the surface, where pressure is minimal. Another is to strengthen the shell so it can resist pressure, the way we build submarines. Humans do it with engineering; cephalopods do it with evolution.

A thicker shell wall relative to its volume can resist more pressure. This is easiest to accomplish if you have a tiny shell with a low volume, so the small size of some shelled cephalopods, both ancient and modern, could be an adaptation to deepwater living. Larger chambers require proportionately thicker shells, which require more time to build. This is a likely explanation for why modern nautiluses take many years to reach modest sizes.

Ammonoids, however, may have found a shortcut to growing strong shells quickly. Instead of thickening the shell wall, they made the seals between chambers incredibly intricate. The sutures of a nautiloid follow simple lines, but in ammonoids the sutures are folded and curled in designs of fractal complexity. Some scientists think that these fancy seals spread out water pressure, leaving no weak spots where the shell would fail.[18]

FIGURE 3.1 Three-dimensional surface renderings of the shells of two species of modern nautiluses (*left*) alongside two species of ancient ammonoids (*right*) showcase the difference in complexity between their chambers. A single chamber of each species is shown beside the full shell to give a clearer view of the septal shapes.

Robert Lemanis, Dieter Korn, Stefan Zachow, Erik Rybacki, and René Hoffman, "The Evolution and Development of Cephalopod Chambers and Their Shape," 2016.

The intricacies aren't haphazard. Their consistency, in fact, makes them a primary tool for paleontologists to describe and identify ammonoid species. In any given animal, each of dozens of septa separating the chambers are all formed the same way, like stacking cups. This is true not only within an animal but within a species. And while each species has its own distinctive pattern, there's a clear relationship between them. Most of the earliest ammonoids displayed fairly simple suture lines, with increasing complexity appearing as evolution progressed.

Perhaps not coincidentally, the earliest ammonoids with the simplest sutures were also some of the tiniest. Increasingly complex sutures could have let ammonoids get away with a thinner shell, so they could grow faster and therefore get bigger without requiring an extended life span.

Another possible benefit of complex sutures could be escaping predation. Just as they could resist water pressure, sutures could resist the pressure from a predator's jaws, making the shell more difficult to crack. Even if the shell failed in one place, the curves of the seals may have prevented the fracture from spreading. Faced with such an ineffective bite, a frustrated shark might decide to go find easier food — and the ammonoid might even have been able to repair such an injury faster than its nautiloid cousins. Many fossils confirm that ancient cephalopods could survive and heal from attacks. Any puncture to the phragmocone itself was lethal, as the siphuncle, the only soft tissue in the phragmocone, can't grow more shell. But if the living chamber was damaged, well, the animal's capable mantle was right there to fix it.

Strengthening the shell and boosting growth rate are compelling explanations for ammonoid sutures, but it's possible that neither was the initial advantage. Some scientists think that when complex sutures first evolved, they might have played a role in gas exchange.

It takes a long time for modern nautiluses to change the liquid-to-gas ratio in their chambers, by either drawing out water with the siphuncle or letting it seep back in. They can do it as they grow, when new chambers are formed, and they can do it to accommodate the change in buoyancy if a piece of shell is broken off. But they don't do it in order to move; to swim up or down, a nautilus just angles its siphon appropriately and jets.

More wrinkled septa might have allowed ammonoids to change their liquid-to-gas ratio, and therefore their buoyancy, more quickly than modern nautiluses. In this scenario, they might even have been able to move like a hot-air balloon, which rises or falls not because of active propulsion but because the balloonist changes the vessel's buoyancy by adding or removing heat.

If complex sutures began this way, then later ammonoids may have co-opted them to provide shell strength. Such shifts in a trait's use over time are fairly common. Feathers, for example, were most likely used for insulation when they first evolved in dinosaurs. It took a long time for evolutionary tinkering to reach the realm of aerodynamics — and even modern birds that use feathers for flight still benefit from the warmth of down. Similarly, while an initial increase in ammonoid suture complexity may have conferred advantages in movement, as time passed ammonoids with more complex sutures began to benefit from their shells' increased ability to resist pressure and predation.

Begging indulgence for a moment of anthropomorphism, I like to imagine that evolution felt about ammonoids the way my daughter feels about Play-Doh. The possibilities are endless, the thrill of creation almost overwhelming.

And then, despite their intricate sutures and prolific spawning and tightly coiled shells, ammonoids were nearly erased from the oceans.

Ninety-Six Percent: The Great Dying

Two hundred and fifty-two million years ago, Earth opened up its guts in a massive case of planetary indigestion and gave everybody hell. We know this because of the element carbon, which is used by all living organisms to build their bodies and by a very small subset of living organisms (scientists) to study Earth's history.

Carbon comes in a heavy version and a light version, both of which are readily available all over the planet. As Dieter Korn explains, "Organisms like to take from the global pool the lighter one to make their soft parts. If there is a lot of life on Earth, the pool is depleted in the lighter carbon."[19] When organisms die and decompose, their carbon is returned to the global pool.

Throughout the Permian, there was a lot of life on Earth. And so the available carbon was light on the light kind and heavy on the heavy kind. Then, about 250 million years ago, the rocks record a sudden influx of light carbon. "This can hardly be explained by natural processes," says Korn. "Something was wrong with the carbon cycle, and the main hypothesis is that there were these big volcanoes in Siberia. An extremely big volcano burnt all the organic material in this area, and this was then the source of the light carbon."

"Extremely big" is a bit of an understatement. The Siberian eruptions are thought to have lasted 100,000 years, and their geologic impacts are easily visible today as around 772,000 square miles (2,000,000 km²) of natural basalt pavement.[20] Obviously, anything living in the floodplain died, but somehow this regional problem expanded into a global mass extinction marking the end of the Permian. Though less celebrated than the one 200 million years later that killed the dinosaurs, it was more devastating; quite sober scientists have given it the dramatic name "the Great Dying." It took a brutal toll on nearly every group, from wiping out 70 percent of vertebrates to exterminating substantial quantities of insects — the only time a mass extinction has affected this incredibly resilient group.

How could a volcano, even a group of volcanoes, even a very big group of volcanoes, have such broad effects? How could it nearly end the ammonoids? Korn, the grand old man of cephalopod paleontology, doesn't know. If he doesn't know, nobody does.

Halfway around the world from the Museum für Naturkunde, Professor Matthew Clapham at the University of California, Santa Cruz, is working to understand the catastrophe. I ask him what life would have been like for those doomed creatures that lived during the end of the Permian. While I understand intellectually that 100,000 years is an eyeblink of geologic time, it's also as long as the entire lineage of modern humans, *Homo sapiens sapiens*. I can't conceive of volcanism on that scale.

FIGURE 3.2 *Cenoceras* is a nautiloid from the Triassic, descended from the extinction-resilient Permian nautiloids. *Franz Anthony*

Clapham explains that volcanoes weren't erupting every day or even every year of the 100,000. He thinks there might have been one big eruption every few hundred or thousand years, each lasting no more than a few years. Some of these individual explosions must have been truly stupendous, way beyond those issuing from Mount St. Helens or Krakatau, because otherwise their impacts on the environment would have been too gradual and the global system would have buffered the change. No cataclysmic change, no mass extinction. Clapham even theorizes that a period as short as a hundred years did most of the damage. "We'll never have the resolution to see it," he says. "But it must have been like that."[21]

If ancient nautiloids shared the lengthy twenty-year life spans of their modern descendants, then individual animals might have witnessed significant disturbances to their environment. As the earth belched out massive amounts of carbon dioxide, the global temperature was cranked up by 11–18 degrees Fahrenheit (6–10°C). Equatorial ocean water might have reached or even exceeded safe hot-tub temperatures. Warm water holds less oxygen than cold water, so oxygen levels plummeted. At the same time, the ocean absorbed excess carbon dioxide from the atmosphere, which led to a chemical reaction that lowered oceanic pH.

According to Clapham, the big unknown is whether the drop in pH really affected marine organisms. If you've read about modern ocean acidification as a result of industrial belches of carbon dioxide, this may seem obvious — bleak predictions of coral and seashells eroding or unable to form in the first place grow closer to reality every year.[22] But that's because in modern times the pH is changing so *fast*. If it changed more slowly, ocean feedback loops could buffer the change and leave plenty of carbonate for shell-building organisms.

If volcanoes at the end of the Permian did manage to drive a change in ocean pH as drastic as the one we're creating today, this might explain why the greatest death of the Great Dying occurred in the sea, where 96 percent of all marine species disappeared. Invertebrates were harder hit than vertebrates, but still a great many sharks and rays were lost. Ammonoids, of course, were devastated.

But nautiloids were not.

"This is something that happened to ammonites versus nautiloids in all the extinction events," notes Korn. "The ammonites were always first to suffer, the nautiloids not."[23] To explain the differing effects on these two superficially similar cephalopod groups, paleontologists often turn to their different reproductive strategies. Something about the environmental changes must have favored the long-lived, large-egged nautiloids, while punishing the short-lived, small-egged ammonoids.

Whatever the reason for their divergent rates of extinction, it would be easy to guess that nautiloids were prepared to take over in the wake of the ammonoids' destruction. And yet a few ammonoids did squeak through . . .

4

The Protean
Shell

Although we've said farewell to the days when cephalopods were the biggest and brawniest sea beasties, in the Mesozoic they found a new kind of evolutionary success: abundance and diversity. Though most people see dinosaurs as the iconic fossils of this era, to many a paleontologist that position is occupied a thousand times over by ammonoids.

The Paleozoic ammonoids we met in the last chapter built some nice shells and had plenty of babies, but it's Mesozoic ammonoids that give the group its fame. All the most curious shapes, all the most staggering abundances, are found in the Mesozoic periods: Triassic, Jurassic, Cretaceous. It is these ammonoids that are most useful as geologic timestamps, and it is these same ammonoids that begin to answer the deep questions of scientists like Peg Yacobucci: How does evolution happen? How do new creatures arise?

Let's begin by looking at the world into which these creatures were born. That great continental conglomeration, the "all-earth" Pangaea, had assembled during the late Paleozoic, so by the time of the Great Dying all of Earth's land was gathered in one gregarious mass. What we now call Australia clung to Antarctica and India, which in turn stuck to Africa and South America, which merged continuously into North America and Eurasia. These landmasses would spend the next 175 million years drifting apart into the continents we're familiar with today.

From a terrestrial point of view, the geologic story of the Mesozoic is the breakup of Pangaea. But from an aquatic viewpoint, the tale is the determined infiltration of that supercontinent by Panthalassa, or "all-ocean"—the enormous expanse of water that once lapped at every coast.

During the Triassic, one wet wedge split Pangaea into north and south halves, creating the Tethys Sea in the process. Next, at the start of the Jurassic, the young and eager Atlantic Ocean pushed its way between North America and the conjoined mass of South America and Africa. As this and other megacontinents split further over the course of the Cretaceous, more and more seaways opened and spread between them. Tethys, the forerunner of all this oceanic activity, was eventually reduced by the incursion of India and Africa into what we know today as the Mediterranean Sea.[1]

Perhaps it was the diversification of the seas that facilitated the diversification of the animals therein. Evolution does not happen in a vacuum; new creatures must adapt to new niches. Abundant, beautiful, and bizarre, Mesozoic ammonoids were shaped by changes in their environment, from stagnant water to rising floods and from tasty new plankton to terrifying new predators.

Rising from the Ashes

After the Great Dying, the next 50 million years continued to be a rather volatile and dangerous time to be alive. Conditions like these, though, can drive evolutionary innovation, and indeed paleontologists describe a "Triassic Explosion" of animal diversity, comparable to the Cambrian in scale. Also like the Cambrian, it happened mostly in the ocean—the enormous part of planet Earth that people tend to forget about. Cephalopods seem to have

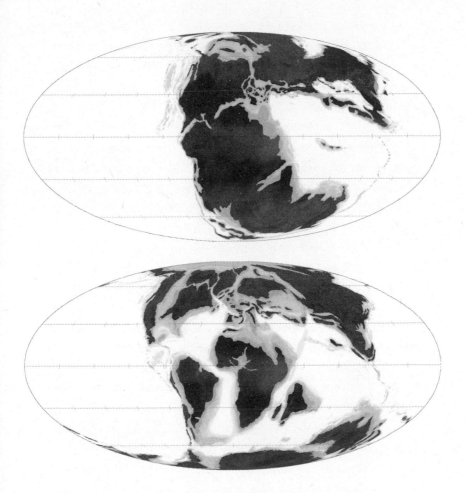

FIGURES 4.1A & 4.1B The shift in continental position
from 180 million years ago (early Jurassic) (*top*) to 80 million
years ago (late Cretaceous) (*bottom*). Note the breakup of
Pangaea, the appearance of the Western Interior Seaway,
and the opening of waterways all around Antarctica.
Global Paleogeography and Tectonics in Deep Time,
© 2016 Colorado Plateau Geosystems, Inc.

been particularly well suited to take advantage of the dynamic environment. They bounced back after each aftershock of the Great Dying, experiencing a relatively low level of competition and predation from vertebrates.

On land, vertebrates were getting quite busy. The Triassic is known for the evolution of the first dinosaurs and mammals, although they were limited in size and abundance, and were preyed upon by enormous proto-amphibians called Temnospondyli. The mostly landlocked environment was dry and barren, a far cry from the lush terrestrial world of the later Mesozoic. But if Pangaea was an arid desert, not so Panthalassa.

Triassic oceans were a hotbed of evolutionary radiation. Warm water ringed the supercontinent and its outlying islands, and there were no ice caps at the poles. The first marine reptiles showed up in the early Triassic, and ammonoids recovered quickly from their near obliteration. Even some of those curious Temnospondyli adapted to life in the sea, which is virtually unheard of among amphibians. The pelagic realm, in short, was doing great.

The seafloor took longer to recover, perhaps because of a persistent lack of oxygen. The volcanic outgassing that had ended the Permian left its signature on the global climate. Ocean water saturated with carbon dioxide and heated by the greenhouse effect was hard for aquatic animals to breathe. Bacteria that thrived in a low-oxygen environment perpetuated the conditions that created it in the first place.

Cephalopods may have been some of the few creatures that could handle the stress, if they were anything like their modern descendants. Not only do modern nautiluses have the unusual ability to breathe easily when carbon dioxide levels are high, but they can also reduce their metabolism in low-oxygen water. And they can pump enormous amounts of water over their gills to maximize the extraction of whatever oxygen there is. Squid have a few tricks up their mantles as well — two species in particular, the Humboldt squid and the vampire squid, are well able to lower their metabolic needs when oxygen is hard to come by.[2]

With this ability appearing in both modern nautiloids and coleoids, it's not unreasonable to guess that at least some ammonoids could deal with low oxygen levels too. We can imagine that when oxygen started to seep back into the ocean in the early Triassic, ancient cephalopods would have been well poised to seize advantage of this precious gas's first availability.[3] Most other animals would have to wait until oxygen levels climbed higher, while throughout the Triassic both ammonoids and nautiloids continued to diversify.

Even for cephalopods, though, the Triassic wasn't a walk in the park. They were knocked back during at least two major and a couple more minor extinction events within it, and a seriously extreme event at the end of it.

The cause of the mass extinction at the end of the Triassic is unknown. Occurring just over 200 million years ago, a scant 50 million years after the Great Dying, could it have been driven by similar volcanism? Flood basalts, similar to those of the end-Permian, have been found aligned with the timing of the end-Triassic extinction. And we know that when volcanoes drive up global temperature and perhaps oceanic acidity, animals face all kinds of trouble. Those with especially sensitive physiologies might be killed outright by the environmental change, but even those that can tolerate the heat and lowered pH must struggle. Marine animals often rely on chemical signals passing through the water in order to find mates, and when the water changes, these signals can go haywire. Confused and unable to reproduce, species may face a poignant "death by celibacy," in the words of the paleontologist David Bond.

Though its causes are uncertain, the impact of the end-Triassic extinction event was dramatic. Killing off most of the dominant predators on land, including Temnospondyli and crocodile cousins, it opened up space for the dinosaurs. In the sea, nautiloids were reduced to nautilids alone, the single narrow lineage that would lead to modern nautiluses. Nautilids puttered along from that day to this, growing slowly, building thick shells, and laying large yolky eggs.

As for ammonoids, so many of them died off at the end of the Triassic that scientists have been challenged to find a single species that crossed the boundary into the Jurassic — though several must have done so, to become progenitors of subsequent generations. Whoever these survivors were, their small eggs and rapid growth immediately aided their evolution into a new wave of ammonoids suited to the new marine landscape.

The global environment had stabilized, so grab some popcorn and settle in to witness the opening of a Jurassic aquarium.

Getting Defensive

If cephalopods thought that fish had been a tough threat, well, fish were nothing compared to the marine reptiles that flourished in the Jurassic and Cretaceous. Ammonoids around the world were now on the menus of three distinct groups of land reptiles that had independently returned to the sea.

First came the ichthyosaurs, "fish lizards" with big propulsive tails, four smallish flippers, pointy heads, and no appreciable necks. They looked, and probably behaved, a lot like dolphins (which are the bane of modern squid). Ichthyosaurs were followed by plesiosaurs with long necks, small heads, and short slender tails that gave them a resemblance to the massive long-necked sauropods on land. Plesiosaurs propelled themselves with four large muscular flippers and probably went after small, slow prey. But then some plesiosaurs grew shorter necks and bigger heads and were given (by paleontologists, millions of years later) species names like *ferox*, suggesting that they were rather more *ferocious* hunters of large prey. The reptiles in this short-necked subgroup of plesiosaurs are called the pliosaurs, in what is surely one of the most unnecessarily confusing bits of nomenclature ever. Fortunately, they can still be recognized as plesiosaurs by their big muscular flippers and slim tail.

Ichthyosaurs had begun to fade away by the Cretaceous, when the third group of reptiles showed up with a reprise of the large propeller tail that ichthyosaurs had sported. These were the mosasaurs. Perhaps these opportunistic creatures were taking advantage of a niche left vacant, or perhaps they outcompeted the remaining ichthyosaurs.

We can be fairly sure that all of these reptilian predators ate cephalopods. Shell remnants have been found in the fossilized stomachs of plesiosaurs and ichthyosaurs,[4] and mosasaur bite marks have been found on fossilized shells. One remarkable shell even bears bite marks from both a larger and a smaller mosasaur, which one scientist took as evidence of a parent instructing its offspring in proper hunting technique.[5] That might seem a little far-fetched — it's at least as likely that two unrelated mosasaurs of different sizes were squabbling over food, one snatching the other's prey.

It's nowhere near as far-fetched, though, as paleontology's most bizarre tale of marine reptiles and cephalopods. Remember Mark McMenamin, the paleontologist who named the Garden of Ediacara? He's notorious within the geology community for "launching very controversial ideas," as De Baets puts it. "He's always a bit overdoing it."[6]

In 2011, McMenamin and his wife, Dianna Schulte McMenamin, announced that they had evidence for a huge, ancient cephalopod they dubbed the "Triassic Kraken." No part of the Kraken itself had fossilized, but they believed that it had preyed on ichthyosaurs and then arranged the dead animals' bones into a *self-portrait*. This interpretation was based on a particular rock formation in Nevada, where the fossilized vertebrae of nine large ichthyosaurs are found in curious double rows that paleontologists have sought to explain in various ways over the years.

Two hundred and fifteen million years ago, landlocked Nevada lay under a warm shallow sea, which was home not only to many swimming reptiles but to a large array of cephalopods as well. Diverse ammonoids and coleoids roamed this Western Interior Seaway, but none were very large — especially not in comparison with the fifty-foot-long ichthyosaurs, which were certainly avid predators of shelled cephalopods.

The McMenamins hypothesized, however, that the Seaway was also home to an unshelled cephalopod so enormous that it could actually take down a fifty-foot ichthyosaur. This "Kraken," they said, must have killed the ichthyosaurs, feasted on their flesh, and then, in a flash of inspiration, arranged the vertebrae of its prey into patterns resembling the rows of suckers on its own enormous arms.

Soft-bodied cephalopods have a notably sparse fossil record. That far more species of ancient octopuses existed than we have fossil evidence for is virtually indisputable. But that at least one of these species was not only many times larger than any known fossil, but many times larger than any modern octopus as well, is a suggestion that strains belief. "It's not science," says De Baets. "I always feel a bit bad if we [paleontologists] are in the news with this kind of story. Someone would think, what are these people doing? Just coming up with crazy ideas all the time?"

Paleontologists don't really need to come up with crazy ideas, at least when it comes to cephalopods. Evolution's got that department completely covered — as we can see from the evolutionary arms race that happened over the course of the Mesozoic.

Three groups of predatory marine reptiles would seem to be more than enough for anybody to contend with. But marine reptiles weren't the only shell breakers, and cephalopods weren't the only victims. The perpetrators of the Mesozoic's widespread shell predation were fish and sharks, crabs and lobsters, and even snails, taking no pity on their kin. They crunched, cracked, drilled, and pried their way into virtually every mollusk shell in the sea. The evolutionary results are recorded in rock, as natural selection drove ongoing advances in armor.

Mollusks built thicker shells. They grew long spines as predator deterrents. They made smaller shell openings, sacrificing their own wiggle room in exchange for a more defensible front door. The mollusk specialist Gary Vermeij has found evidence in every kind of fossil mollusk that the Mesozoic was a time of intense defense. The pattern is so dramatic that he dubbed it the Marine Mesozoic Revolution.[7]

Like all their fellow mollusks, cephalopods adapted at top speed. Ammonoids changed their shell shapes so that they could hide their soft parts deeper inside, and they developed complex structures on the outside. Paleontologists refer to these various shell protrusions as ornaments, but it's unlikely that their purpose was purely ornamental. Over geologic time, ammonoid shells display more and more spines as the incidence of bite marks increases, indicating that the former probably evolved to prevent the latter.

Evolution of shell ornaments, in fact, has given Peg Yacobucci a key to the whole wild pageant of ammonoid diversity.

The Recipe for Change

As you can see from the sample platter in figure 4.2, ammonoids evolved a stupendous array of shells. But many of the features that make species look so different from each other are really the same feature, expressed at different ages. A few genetic twiddles to the developmental controls could be all it takes to fill up the ocean with new species — as long as the ocean has enough niches to spare.

For example, around the middle of the Cretaceous, the ancestral ammonoid genus *Plesiacanthoceras* gave rise to the younger genus

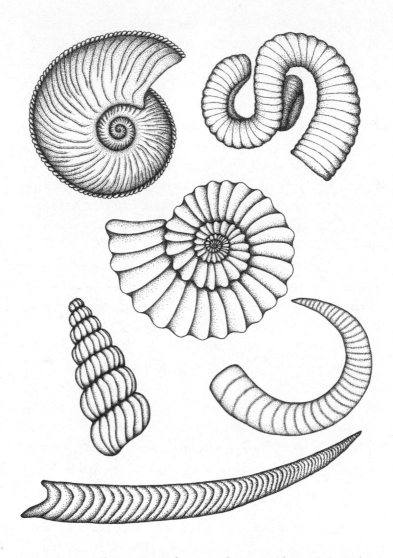

FIGURE 4.2 The enormous diversity of ammonoids is only hinted at
by this display. *Top left*: oxycone with keel; *top right*: *Nipponites* (bizarro
knot); *center*: serpenticone with ribs; *lower left*: *Turrilites* (coiled like
a snail); *lower right*: cyrtocone; *bottom*: *Baculites* (straight shell).
C. A. Clark

Metoicoceras. The names are a mouthful but, again, *ceras* just refers to the horn shape most ammonoid shells have, so we can drop it. *Plesiacantho* means "old and spiny," both useful features to remember about these ammonoids. *Metoico* is a bit less pragmatic and more poetic; it means "wanderer." The creatures were so named because after originating in North America they spread around the globe.

We'll go ahead and follow Yacobucci's lead in shortening the names affectionately to Plesi (Plee-zee) and Metoico (Meh-toy-ko). Flip to figure 2.4, the grand evolutionary history of all cephalopods, and you can see Plesi and Metoico holding tentacles across the Ammonitida branch of the family tree. Here's Yacobucci introducing them: "Plesi has big beautiful spines, and Metoico has ribs; it has these pretty, low, rounded ribs. You'd look at these guys and think they're completely different shells. But as young juveniles their shells are identical. You couldn't tell them apart."[8]

If ammonoids were like most other animals, it would be really tough to figure that out. You'd have fossil juveniles and fossil adults, and how would you know which juveniles grew to be which adults? But as we learned, the beauty of ammonoids is that each adult carries its juvenile form along with it. "So you can follow it from hatching to adulthood on a single shell," says Yacobucci.

Having done this many times, she notices how changes that seem impossible to enact at the level of an adult animal become simple to understand in the context of a developing animal. It's classic evo-devo, as seen in all kinds of animals from sea urchins to whales. "What drives changes is developmental programming," explains Yacobucci. "If you want to get rid of back legs, you just have to shut off the gene that tells the embryo to make legs, and then you get a whale." Similarly, she suggests, "Maybe the way we can make ammonites quickly is highly plastic — monkey with one little gene and now you have a spiny one instead of a ribby one."

The fossils of Metoico and Plesi certainly lend themselves to this interpretation. The babies of both genera have a kind of ornament that's different from those displayed by either parent: they have bumpy ribs. "And as they grow, in Plesi the ribs get suppressed and these little bumps on the ribs get longer and longer. In Metoico the opposite, bumps are suppressed," says Yacobucci. So both kinds have the necessary genetic instructions for growing ribs and for growing spines, and the babies follow both sets of instructions. As they grow up, a control switch gets flipped. Plesi stops growing ribs and grows spines; Metoico stops growing bumps and grows smooth ribs. Without being able to test the genetics of the creature, it's as close as we can get to seeing a simple change in developmental controls lead to evolutionary divergence in the fossil record.[9]

There's a problem, though. If the new ribbed forms stay in the same place as the spiny forms and keep doing the same things — in particular, if they keep reproducing with the spiny forms — there's no opportunity for divergence. Some kind of separation is needed to cement evolutionary novelty in place.

Such separations are a frequent feature of life's history, and they continue to occur around the world today. A group of insects might be blown in a hurricane from one island to another, or a few unusually adventurous fish could swim from one lake to another. On the new island or in the new lake, the founding population has a chance to diverge, not only from the population on the old island or lake, but within itself. The founding animals can spread out and specialize in empty niches. One group of fish could adapt to life on the sand, another to rocks; as they isolate themselves and breed only with other local fish, they accumulate evolutionary changes. After a while, the sand species and the rock species have different colors. Then, different shapes.

Ammonoids had an excellent opportunity to colonize and diversify into a new habitat like this in the mid-Cretaceous, when the heartland of what is now the United States and Canada was inundated with rising seas. This was the same Western Interior

Seaway that the hypothetical Kraken shared with truly monstrous ichthyosaurs, and it can be seen bisecting North America in the second map of figure 4.1. The Western Interior Seaway is a tremendous boon to paleontologists today, as it left behind an abundance of marine fossils in rocks that are relatively easy to access. The Seaway was also a tremendous boon to ammonoid diversity 90 million years ago, as it had an irregular seafloor and plenty of nutrients washing in from rivers, making for a highly variable habitat in which many species could specialize.[10]

Not only changes to the physical habitat, but changes in the surrounding ecology, may have facilitated Cretaceous diversification. Back in the Paleozoic, the oceanic menu wasn't very long —only a page or two of options. Sure, there were some plankton blooms, as we saw in the Devonian with the influx of terrestrial nutrients that led to ammonoids hatching tiny plankton-gobbling babies. But plankton evolution in the Mesozoic took many steps further, filling in the marine food web with all kinds of new species. The menu expanded to nearly book size, and the increase in quantity and diversity of nutrients fueled comparable increases in the quantity and diversity of ammonoids.

Of course, ammonoids were not the only animals or even the only cephalopods to live in this world of rising seas and nutrients. Why did they evolve so much faster than their close cousins, the nautiloids? Back in the Paleozoic, nautiloids and ammonoids were probably doing much the same thing—bobbing around as smallish spherical shells. Neither group was tremendously abundant, so they could overlap in that niche. Then, in the Mesozoic, ammonoids became superabundant and superdiverse, and nautiloids, well, didn't. Perhaps it's all down to that developmental plasticity; perhaps the ammonoid genome was unusually pliable.

The malleability of ammonoids allowed them to evolve a variety of defensive structures, like Plesi's spines. But the range of ammonoid forms also encompasses a wide array of ornaments that are less obviously off-putting. What about Metoico's ribs?

A Shell Is Such a Drag

Like Yacobucci, the paleontologist Kathleen Ritterbush, of the University of Utah, began her career fascinated with mass extinctions. She wanted to know why ammonoids were devastated (when they were devastated) and why they survived (when they survived) and why they eventually disappeared altogether. For her PhD she studied the end-Triassic extinction, which was perhaps the second-worst time ammonoids have ever had. But then Ritterbush was struck by the Jurassic flowering of ammonoids — particularly one group known as the psilocerids.

"They're cosmopolitan, they're abundant, and they're large. Some of them were a half meter," she says. "Whatever it is they're doing, they're doing it right. At first they're smooth. Within two million years after the extinction event, they *all* have ribs. And I'm talking legit ribs." To make sure I have gotten the point, she offers another adjective — *"ridiculous* ribs" — and suggests, "When something is that glaring of a signal in the fossil record, it had to be good for something."[11]

But the received wisdom in the field of ammonoid paleontology was that ornaments like ribs simply showed that the animals bearing them were terrible at swimming. The historical habit of basing ammonoid biology on living nautiluses reinforced the notion of ammonoids as uninspired swimmers; as Ritterbush points out, "Current nautilus are spectacularly crappy at swimming."

Frankly, jet propulsion is not a great way for a shell-bound animal to move. Coleoids like squid and octopuses can inflate their mantle like a balloon, then squeeze it like a fist; this is called *mantle pumping*. Nautiluses can't inflate because they're stuck in a rigid container, so they have to use a less efficient technique called *shell pumping*, wherein they yank the whole body back into the body chamber, forcing water out (see figure 2.2). Alternatively, nautiluses can ripple the edges of the siphon to create a slow, continuous flow of water that both aerates their gills and propels them gently through the water. While elegant, this technique doesn't get them anywhere in a hurry.

It might seem that ammonoids would all have been restricted to the same kind of inefficient shell pumping as nautiluses, and many probably were. But others may have evolved tricks to get around this limitation. Some intriguing fossils indicate that several ammonoid species may even have experimented with bringing the shell inside the mantle, like our old Silurian pal *Sphooceras*.[12] Enwrapping the shell with soft tissue could reduce drag, while parts of the shell could then become more of an internal skeleton. Projections around the shell's opening, for example, might have supported a large muscular siphon for active swimming. And even for the majority of ammonoids with wholly external shells, ornaments could have significantly mitigated drag. After all, the dimples on a golf ball make it fly faster.

Alas, we can't dredge live animals out of the rocks and watch them swim. The best we can do is test ammonoid models. Scientists have been building such models for decades, relying heavily on estimation and guesswork. In grad school in the 1990s, Yacobucci made some models herself. Looking back now, she laughs. "It was me with some clay and a picture of an ammonite."[13]

Recent technological advances, such as three-dimensional scanners of the same quality as those used for medical tests, have introduced a level of precision not previously possible. For example, one of the most important parameters in calculating the force of a jet, and therefore an ammonoid's swimming speed, is the volume inside the living chamber. That's incredibly difficult to estimate for ammonoids with complex sutures (which is to say, most ammonoids). But 3D scans mean we don't have to estimate. We can measure.

Such scans become even more powerful in conjunction with the rapidly advancing field of 3D printing. Ritterbush is in the midst of creating a laboratory setup that would allow her to pick any fossil, scan it, and print it. Just as premium golf balls are constantly tested for speed and spin, Ritterbush's goal is to plop ammonoids of every shape into a water tank for testing. Her succinct description of the project is "to look at the shell as just a pain in the ass." There's no getting around the fact that a shell is fantastically annoying to an animal trying to travel through the water, and with her model setup, she hopes to illuminate all the ways evolution might have dealt with that challenge.

Once they have a library of scanned fossils, Ritterbush and her students can tweak any given parameter and ask: What if the whole shell were wider? What if just the opening were wider? Narrower? What if it had a sharper keel, or no keel? (Keels are raised ridges along the outside of some ammonoid shells, and their very name reveals paleontology's assumption that they provide stability during fast swimming, just as the protrusion humans build along the bottom of a boat keeps it from tipping over. But that assumption has never been tested.) And finally, what we're all wondering: Are those ridiculous ribs a hydrodynamic perk or a liability?

Ritterbush's project offers a much-craved opportunity to examine in the tangible world an interpretive scheme of ammonoid form and function: a tidy little triangle outlined by the late German paleontologist Gerd Westermann and named by Ritterbush in his honor *Westermann Morphospace*.[14] Westermann had observed that for all their diversity, most ammonoid shapes varied between three simple types: slim disks named *oxycones* (from the Greek root for "sharp"); loosely coiled *serpenticones*, which look like snakes; and fat globular shells called *spherocones*, which look like, you guessed it, spheres. These shapes emerge as the shell grows, due to variation in the width of new coils or the degree to which they overlap old ones.

Serpenticone   Spherocone

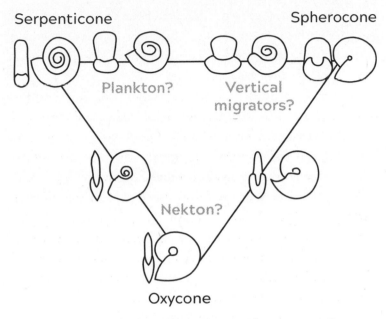

Plankton?   Vertical
migrators?

Nekton?

Oxycone

FIGURE 4.3 Westermann Morphospace illustrates prevailing
hypotheses about ammonoid behavior based on shell shape.
Kathleen Ritterbush

The quickest and nimblest were probably the oxycones, throw-
ing themselves through the water like discuses. Paleontologists
suspect they were active predators, jetting after their prey and cap-
turing it alive. They may also have migrated long distances, like
many of today's whales and seabirds.

Both the loose serpenticones and the globular spherocones are
thought to have experienced too much drag to move quickly. Instead of hunting, they might have used their arms and jaws to sieve
the water, consuming whatever tiny particles they happened across.
Westermann interpreted the serpenticones as drifting plankton,
perhaps akin to the early slow-swimming cephalopods of the Ordovician, and the spherocones as vertical migrators.[15]

Vertical migration is a habit that lots of marine animals share
today. So many animals live in deep water during the day and

migrate to the surface at night that their bodies form a layer dense enough to reflect sonar. During World War II, ships sometimes mistook this layer for the seafloor; crews must have been somewhat distressed when it began to rise.

Eventually scientists discovered that this "phantom bottom" was an incredibly thick aggregation of fish, jellies, and shrimp, all of which had figured out that sunlight does two things at once: grow food and make it hard to avoid predators. So they visit the sun-stimulated surface buffet under the protective cover of night, then hide in the dark depths during the day. Numerous modern cephalopods, especially squid living in the open ocean, participate in these daily migrations. As carnivores, they certainly don't eat the algae that bloom at the surface; instead they snack on the grazers—either their fellow migrators or creatures that are too tiny to migrate and are thus stuck at the surface.

A number of Mesozoic ammonoids likely followed this up-and-down lifestyle. The spherocones might have been joined in their daily routine by some ammonoids so peculiar they have no place on Westermann's triangle, existing in a totally separate morphospace of their own: the heteromorphs, or "other shapes." Among the multitude of heteromorph oddities are a number of helices shaped like soft-serve ice cream. They would have been challenged to swim in straight lines, but migrating up and down in slow spirals would have been an efficient way to extract every particle of food from the surrounding water. Monks and Palmer give this poetic account in *Ammonites*: "Imagine a Cretaceous evening scene a few miles offshore at dusk. At a depth of 100 meters or so, vast shoals of helical ammonites . . . are slowly corkscrewing their way upwards. . . . For tens of millions of years, these elegant, pirouetting predators must have been a very distinctive and beautiful part of the marine realm."[16]

The Other Shapes

Unlike streamlined swimmers or delicate drifters, heteromorphs just seem like Ammon playing a trick on us. In addition to helical ice-cream cones, heteromorphs grew in the shape of large aquatic paper clips and in bewilderingly recurved knots. For now, scientists have gathered them all into one group (as you can see in figure 2.4), but it's debatable whether they're actually related to each other or whether this is merely a label of convenience. Far more hotly debated is the question of what the living animals could possibly have been doing with all these bizarre shapes. Theories abound, some more plausible than others. In recent years, evidence has mounted to suggest that the two most abundant groups of heteromorphs did not swim fast, nor yet migrate vertically— instead, they hovered near the seafloor, scavenging or filtering odds and ends from the water and from the ground.

The most abundant kind of heteromorphs were called baculites. These creatures didn't live in a coil, a helix, or a knot. The shell of a baculite was quite simply straight—an unexpected throwback to the early days of cephalopods. For decades at least, paleontologists have puzzled over how this straight-shelled ammonoid would have looked in the water. Did it sit vertically, which might suggest it was another migrant? Or horizontally, making it more like a fast-swimming squid? It is curious that ammonoids, so committed to curling, did evolve one straight lineage, and it's tempting to suppose that this might have been an evolutionary result of competition with the coleoids. Finally, a geologic revelation in 2012 led to a new perspective on baculites.

One of the largest shale formations in North America, the Pierre Shale, contains fossilized examples of entire ecosystems known as *methane seeps*. They formed yet another component of that broad Western Interior Seaway, and they probably functioned much like the ones in our modern oceans. Methane seeps as we know them today begin with methane and hydrogen sulfide gas bubbling up from underground. These chemicals attract gas-hungry bacteria that attract grazers, which then draw larger pred-ators — octopuses today, ammonoids in the Cretaceous. Perhaps due to the unusual local chemistry, animal fossils within seeps are even better preserved than those in surrounding shales.

Unfortunately, by the time paleontologists evolved, most of the methane seeps in the Pierre Shale had already been exposed — and worn away. Then at last, in one of the lucky breaks that the science of dead things depends on, a landslide revealed an untouched seep.

Among the rejoicing geologists was the eminent ammonoid paleontologist Neil Landman. As I learned at a 2016 awards re-ception of the Paleontological Research Institution, Landman is a curator, curator-in-charge, *and* professor at the American Mu-seum of Natural History in New York, and, as a speaker noted, he "has done more than any other in bringing ammonoids to life." His work on the methane seep ammonoids is one more entry in the long list of his contributions.[17]

Although ammonoids had been found before at methane seeps, they'd always been thought of as visitors, stopping for a quick bite of methanotrophic-bacteria-farming clam chowder on their way through town. Landman and his colleagues used the chemical composition of ammonoid shells — including baculites — from this new seep to show that they spent their whole lives there, wal-lowing in the gassy water and munching on abundant clouds of plankton.

A few years later, a postdoctoral researcher with Landman published evidence of sedentary baculites at yet another site, also within the Western Interior Seaway. Jocelyn Sessa tackled an as-semblage of fossils at a spot in Mississippi called the Owl Creek Formation that comprised several species of ammonoids along with a variety of other organisms. The beauty of this collection was that the habits of those other organisms were already well known: the clams and snails must have lived on the sea bottom, and there were two kinds of tiny shells called foraminiferans, one benthic and one planktonic. So Sessa could compare the chemistry of the ammonoid shells with that of the other shells and come up with a solid idea of where the different ammonoid species must have lived.[18]

She found that while one group of ammonoids lived way up in the water column, most likely swimming free, two other groups of ammonoids — including baculites — lived with the benthic creatures near the seafloor, perhaps even eating those snails and clams. Because of their buoyant shells, they wouldn't have actually crawled on the ground (no more than would the ones at methane seeps). Instead, these animals were like birthday balloons a few days after the party. No longer bobbing at the ceiling, they drift around the room at eye level, their strings trailing along the floor. Imagine that each string is several tentacles, rummaging through the party debris for tasty morsels, and the round balloon is instead a tapering cone, and you've got a room full of baculites.

The second ammonoid group that Sessa found in close association with the seafloor was also a type of heteromorph. Called *scaphites*, these are the ones that grew their shells into paper-clip-type hooks, and except for baculites they were the most abundant heteromorphs. For decades, everyone thought that they must have sat vertically in the water, with the living chamber opening upward. This would not have allowed them to swim or hunt very effectively, so they would have bobbed along in the plankton, grabbing whatever bits they could reach.

The upward-pointing model was based on physical calculations of a soft body that filled the entire living chamber. In 1998, Neale Monks pointed out that this needn't necessarily be the case.[19] Scaphites could have built living chambers much larger than their bodies and moved around inside them, "rather like a small octopus with a mobile burrow or cave." This motion would have dramatically affected the shell's balance and orientation. If the ammonoid slid toward the shell's opening, it would tilt downward and the tentacles could snatch up food from the seafloor. If the ammonoid sensed predators, it could retreat deep within the shell and the opening would rock away from the threat.

This intriguing and rather odd idea has not been broadly accepted — most paleontologists seem to agree that ammonoid soft bodies must have filled up their final chambers. But it wasn't the oddest idea that would be proposed for scaphites. That came in 2014 from Alexander Arkhipkin, widely known as Sasha, an expatriate Russian working for the government of the Falkland Islands as a fisheries scientist.[20] "I was on the beach here in the Falklands and we have a lot of kelp which is stranded on the beach," he recounts. "And sometimes you can see trees, you know, with very thick branches, and I thought, for goodness' sake, if [scaphites] were hooked, they might be hooked to something like that."[21]

The scaphite hook grows only with the final, adult living chamber; juveniles have more conventional forms. So Arkhipkin envisioned a life cycle in which juveniles swam or drifted around like the majority of other ammonoids. When they were ready to settle down and spawn, they grabbed onto giant seaweeds and grew the hook to attach themselves — actually *functioning* like a paper clip, not merely looking like one. It's a drastic step; once attached, they would never have been able to unhook and might not have even been able to eat. Still, other animals have been known to live peripatetic lives when young, then lock themselves down to reproduce. (You may be able to think of an example in your own family.) And modern octopus mothers are well known for eschewing food once they start to brood their eggs.

As with any provocative scientific paper, a rebuttal followed Arkhipkin's initial publication. Landman and several colleagues dismantled his theory with physical evidence (Arkhipkin had suggested that the irregular wear marks on certain scaphite shells

FIGURE 4.4 The paper-clip theory has been greeted
with skepticism by paleontologists who think it's more likely
that creatures like the one shown here had very short arms
and lived a sedentary filter-feeding lifestyle.
Mariah Slovacek and Neil Landman,
American Museum of Natural History, New York

could have come from rubbing kelp, but pristine fossils show no
such marks), or a lack of evidence (scaphites are never found with
kelplike fossils), and plain logic (if both males and females were attached, as Arkhipkin proposed, they would be challenged to cop-
ulate).[22] Arkhipkin fired back a response titled "If Not Getting
Hooked, Why Make One?"[23]

But Landman had answered that question in 2012, carefully re-
constructing his own vision of these puzzling creatures. He noted
that their shells already show one of the classic antipredator signs
of the Marine Mesozoic Revolution: constriction of the opening, which makes it harder for predators to reach in and grab the

tasty meat. Growing the adult shell into a hook, Landman theorized, would complement this defensive adaptation by tucking the shrunken front door behind a curve — rendering it essentially inaccessible to predators.

Obviously, it would have been very difficult for an animal that looked like a modern squid to poke out of such a hidden opening and do squidlike things. Muscular arms wouldn't have had any room to grab prey. A strong siphon wouldn't have been able to direct the jet to change swimming direction. Consequently, Landman concluded that scaphites probably *didn't* have muscular arms or a strong siphon. They didn't need them, because they hugged the seafloor, as Sessa's work in Mississippi showed. They would have used rather delicate arms, perhaps even an arm web, to eat . . . something. But what?

Well, that depends on the interpretation of a structure that, in Landman's words, "has been under debate for the last 150 years."[24]

The Mouth That Was Mistaken for a Door

Modern cephalopods have beaks rather like hawks, with separate upper and lower halves, sharp enough to suit their predatory lifestyle. Squid and octopus beaks are made of chitin, a stiff composite of linked sugars and nitrogen. Nautiluses, by contrast, have solid beaks made with calcium like our bones. These two kinds of jaws haven't changed for a while; fossilized beaks from ancient coleoids and nautiloids look much the same as the modern ones.

Ammonoids, meanwhile, extended their tremendous diversity of form and habit to their jaws. Some, like coleoids, built them with chitin; others, like nautiloids, built theirs with calcium. And some ammonoids expanded, flattened, and modified their lower jaw almost beyond recognition.

These strange structures got their own special name, "aptychi" (singular "aptychus"), before anyone knew what they really were.

They're much larger than normal beaks and look more like clam shells than like anything a cephalopod might use to eat dinner. Paleontologists in the nineteenth century thought that aptychi might actually be remnants of clams (or barnacles, worms, even birds!) that the ammonoid had eaten, rather than a part of the ammonoid itself. It didn't help that aptychi were often separated from ammonoid shells once the muscles holding them in place decayed. However, by the 1930s enough complete fossils had surfaced to convince scientists that aptychi were part of the ammonoid body.

A new set of creative theories proliferated. Some researchers thought aptychi were large enough to cover the opening of ammonoid shells, shutting out predators and nosy relatives. Snails have doors like this, called *opercula*, which you can see by turning your garden variety upside down. No living cephalopods have opercula, though the leathery hood of a nautilus serves a similar purpose. Other scientists suggested that aptychi might protect specific organs, like gills or ovaries, the way our rib cage protects our lungs and heart. Someone even raised the possibility that aptychi were the shells of parasitic males living inside females. (Not the wackiest idea—there are animals that do this, like deep-sea anglerfish.)

It wasn't until the 1970s that paleontologists had amassed enough evidence to be certain that aptychi were modified lower jaws.[25] Once aptychi are observed in their proper place in a well-preserved ammonoid, it becomes difficult to see them as anything else. A normal upper jaw sits on top of the aptychus, and the structure of the aptychus is clearly derived from a more typical lower jaw.

Although aptychi are far larger than the beaks of living cephalopods, that doesn't mean the ammonoids that wielded them were terrifying predators. After all, *squid* are terrifying predators, yet their beaks are much smaller relative to their body size than ammonoid aptychi. Generations of scientists have produced generations of ideas about how such large mandibles might have achieved caloric intake. Though they might have simply bitten and chewed food like any other jaws, it can be difficult to envision such unusual structures operating in such an ordinary way. The aptychus is so much larger than the upper jaw, and what's even stranger, it comes in two parts. These parts were almost certainly joined by soft tissue in some way. If the tissue was flexible, perhaps the aptychus could have been used as a filter. An ammonoid might have used such an aptychus like a baleen whale uses its enormous jaws, scooping up a mouthful of ocean and then pressing out the water while retaining all the tiny edibles.

Neil Landman prefers a plankton-eating perspective in which the aptychus acts more like a funnel. Delicate arms or an arm web could have directed water flow into this funnel-mouth, and at the back of it a radula like a conveyor belt would have caught small creatures and moved them down the gullet.

However the aptychus might have been used for feeding—and it could have been used in different ways by different species— the early suggestions weren't entirely off base, either. Ammonoids could have adapted their aptychi to nonfeeding uses, and contemporary paleontology favors the idea that aptychi served more than one purpose, like an ammonoid multi-tool. Aptychi could even have helped ammonoids swim. Writing about aptychus function in 2014, Horacio Parent of the Universidad Nacional de Rosario in Argentina, along with Westermann and the American paleontologist John Chamberlain, rattled off this list of previous proposals: "lower mandible, protection of gonads of females, protective operculum, ballasting, flushing benthic prey, filtering microfauna and pump for jet propulsion." Nothing daunted, they proposed an eighth: stabilizing ballast during swimming.[26]

A number of ammonoids might have been rather unsteady swimmers. With the siphon protruding well below the center of the shell, each jet would rock the animal around its axis. Pitching back and forth as you swam would be not only disorienting, but highly inconvenient if you wanted to capture prey or scavenge detritus. Your movement could be stabilized, though, if you could just slide a heavy aptychus out of your shell aperture to steady yourself, rather like a tightrope walker's pole. It makes sense . . . right?

If you're having trouble visualizing all of this, you're not alone. Ammonoid paleontologists have all kinds of sympathy for you. After all, most other animal jaws we know of can be figured out from the hard parts alone. Grab a dinosaur skull, and you can articulate its mouth without too much trouble. But the hard parts of cephalopod jaws are buried in a mass of muscle that gives the jaw its shape — muscle that has never yet been found preserved in any fossil. Ammonoid mouths are still a "mystery," according to Isabelle Kruta, a paleontologist in Paris. "This is why it's exciting to study these structures!"[27]

In 2011, Kruta published a paper in the top-tier journal *Science* doing exactly that.

The Tongue That Isn't a Tongue

For a long time, a fossil radula preserved inside an ammonoid's shell could be seen only if the fossil-containing rock happened to be broken in just the right way. Even then, the details were not usually very good. With new tools, however, we can visualize the ammonoid radula in enough detail to compare it with all other molluscan radulas. Modern mollusks have many kinds of radulas specialized for eating many kinds of things, creating a sort of anatomical dictionary in which the shape of an ammonoid radula could be looked up. This is the dream — just as Ritterbush hopes to study shells closely enough to understand exactly how ammonoids swam, so does Isabelle Kruta hope to study mouths, especially radulas, closely enough to understand exactly how ammonoids ate.

As an undergraduate in Italy, Kruta loved paleontology, so when she decided to do an internship abroad she thought immediately of the American Museum of Natural History. She contacted Neil Landman, who invited her to come work on nautiluses, then ammonoids. She returned to Europe to pursue a PhD in Paris, and it was with both Landman and her French colleagues that she published her groundbreaking research in *Science*: a 3D reconstruction of a baculite radula as it was placed and used inside the aptychus jaw.[28]

The key was tomography, a technique that had been used for decades in paleontology—but only on vertebrates. Because tomographic imaging was developed to look inside human bodies, paleontologists initially thought of applying it only to fossils with bones. Kruta was the first to look at fossil cephalopods with computed tomography, the same kind of CT scan that doctors use to detect tumors and other medical conditions. In computed tomography, a series of 2D pictures are taken as "slices" through the structure—in this case, an ammonoid mouth. A computer then knits the slices together into a 3D view, offering scientists a way to see inside fossils without having to crack them open. (Breaking fossils has always been a risky venture, fraught with the danger of damaging the very structures one wants to see.)

What Kruta found in the ammonoid fossils was an unfoldable radula covered with delicate, comblike teeth. Overall, in shape, it resembles the radulas of modern sea snails that feed on plankton. The 3D scan even revealed little fragments of plankton stuck within the ammonoid's mouth. Such evidence makes a strong case that these baculites—and quite possibly all ammonoids with aptychi—ate plankton.

That said, it's still difficult to envision exactly *how*. Like the paper-clip scaphites, these baculites had very narrow apertures, although the interior of the shell was quite roomy. As we've learned, restricted openings were a common defensive adaptation in the

Mesozoic and may have saved many an ammonoid from death by reptile jaws. But how could the baculite feed itself through such an opening? Landman's vision of a funnel that combines delicate arms with the large aptychus is one possibility. Another is that these ammonoids created some kind of mucus web, like ancient undersea spiders, to catch and entangle any little creature the current swept by.[29] Or maybe they used long slender arms to reach out and grab plankton one at a time—a tiny shrimp here, a little snail there.

However the ammonoids did it, Jakob Vinther for one thinks that eating plankton might be the fundamental explanation for most of ammonoid evolution. Although their initial appearance and radiation in the Devonian may have been spurred by early fish, he opines that they soon adapted to fill entirely different niches, and it shows in their shells. Vinther is among those who consider shell ornaments to be indicative of poor swimming abilities. "Clearly these guys were not actively swimming around to do anything sensible," he says of the diversity of form in latter-day ammonoids. "They were doing their own thing, being these drifting plankton eaters. That's a completely extinct mode of life that we don't see today in any cephalopod."[30]

Indeed, Mesozoic evolution spun the three threads of cephalopod history in wildly different directions: ammonoids followed an action adventure, packed with fast-paced speciation and extinction, death and survival in equal measure. Meanwhile, nautiloids puttered along without much obvious change, characters in a contemplative tale. And coleoids, as we're about to see, displayed one blockbuster success while quietly cooking up another one in the background.

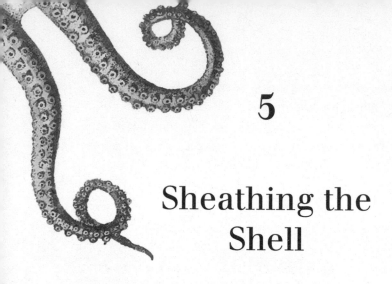

5

Sheathing the Shell

We've seen the influence of fish on cephalopod evolution since the days of the first piscine progenitors, and the convergence of the two groups reached its zenith with coleoids—the group that would spawn modern-day squid, octopuses, and cuttlefish. Coleoids became, in essence, the invertebrate version of fish: streamlined swimming predators with a fast metabolism and a gregarious inclination to shoal. In terms of ecology and behavior, coleoids are far more like vertebrates than they are like any other boneless being, be it clam or worm or starfish or crab.[1]

It was shell reduction that allowed this convergence. Not coiling or truncation this time, but full internalization. Bringing the shell entirely inside the body let coleoids level up in speed and efficiency, abilities equally useful for escaping predators and catching prey.

The first internal shells appeared in the Carboniferous, "only" 50 million years after the Devonian origin of ammonoids. But for the long stretch of Paleozoic time between this origin and the Great Dying, coleoid abundance and diversity were limited. Perhaps the coleoids couldn't yet keep up with their coiled ammonoid cousins.

FIGURE 5.1 This beautiful fossil of the coleoid
Phragmoteuthis conocauda shows clearly the squid-shaped
body with substantial internal shell, the equal-sized arms
(no specialized tentacles), and dramatic arm hooks.
Diego Sala

In the Triassic, they came into their own. Named belemnites after the Greek word for "dart" because of their streamlined shape, these coleoids grew a straight internal chambered shell along with a solid shell guard to serve as a counterweight. On the outside, they looked rather like squid, with two fins and ten hook-covered arms.

Belemnites diversified wildly over the next hundred million years, playing the same keystone role in the Mesozoic oceans as today's squid play in our Cenozoic oceans and as ammonoids had already begun to play in Paleozoic oceans (continuing right up until their extinction): keen predators of smaller things, abundant prey for bigger things. The guts of fossilized marine reptiles are packed with belemnites, and it's been theorized that some prehistoric sharks even died from the overconsumption of heavy belemnite shell guards.[2]

Though belemnites dominate the Mesozoic history of coleoids, during the course of their evolution at least two separate lineages branched off from the coleoid stem: the ancestors of squid and the ancestors of octopuses. Both lineages began with substantial internal shells — as strange as it is to contemplate an octopus with an internal shell — and then both independently reduced this hidden armor. Multiple and varied losses of the shell suggest there was a strong evolutionary pressure to reduce, reduce, reduce. Whence such pressure?

The Mesozoic seas were thick with threats, both predatory and competitive. Coleoids were hunted by the same marine reptiles that attacked ammonoids, and even flying reptiles tried to scoop them from the sea. Meanwhile, bony fish radiated into recognizable ancestors of eels and minnows, salmon and smelt, tuna and goldfish. For both modern fish and modern coleoids, versatility became the name of the game.

What Makes a Coleoid?

Many of the characteristics we consider distinctively coleoid, such as suckers, arm hooks, and ink, begin to show up in the fossil record by the Triassic. These evolutionary innovations could be related to shell loss. As coleoids began to swim faster, they could chase faster prey, which would be more easily captured and contained with the aid of suckers or hooks. But without the shell they were vulnerable, so a new defensive tool arose: ink. Never seen in nautiloids or ammonoids, ink is often preserved in coleoid fossils, thanks to the stability of the pigment melanin. Fossilized belemnite ink was first discovered by English paleontologist Mary Anning in 1826. Her friend and fellow fossil hunter Elizabeth Philpot reconstituted the ink to draw ichthyosaurs, beginning a trend of fossil ink illustration that continues today.[3]

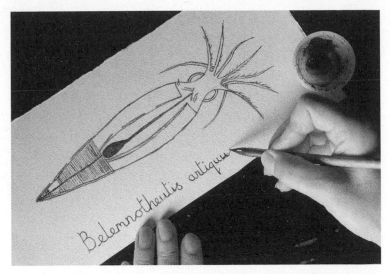

FIGURE 5.2   In 2008, a fossil of the belemnite *Belemnoteuthis antiquus*
was recovered in such good condition that its ink could be
reconstituted and used to illustrate it.

BNPS.CO.UK

But melanin granules in cephalopod ink sacs have turned out to
be far more than a novelty art supply. Their discovery led Jakob Vinther to wonder whether melanin could be found in other fossils —
like dinosaur feathers. It could, and eventually Vinther published
definitive evidence of dinosaur coloration, including a species with
black-and-white-banded wings and reddish head feathers.[4] Despite
the fascination inherent in this work, he has remained actively attached to cephalopods. When I interviewed him, he happened to
have one of the oldest fossilized ink sacs sitting in his office, from "a
really cute little coleoid" of the Carboniferous, about 300 million
years ago. It had a pair of fins, and it had ten arms.[5]

   Enough coleoid fossils like this have been found to substanti-
ate the embryological evidence that a set of ten arms is the ancestral condition — even though nobody kept it to the present day.
Squid modified their *fourth* pair of arms into tentacles; octopuses

modified and eventually lost their *second* pair of arms. This is yet another case of convergent evolution, like nautiloids and ammonoids arriving separately at the coiled shell.

Suckers, on the other hand, are generally thought to have evolved only once, although they've developed to look quite different in modern squid and octopuses. Octopus suckers are flexible and versatile; they can grab and manipulate small objects in addition to suctioning onto larger ones. Squid suckers are more rigid but their suction is much stronger; they sit on stalks like umbrellas blown inside out, and they often contain little rings of teeth as hard as fingernails.

Some species of squid have no suckers at all, instead lining their arms and tentacular clubs with hooks. The most famous of these is the colossal squid, whose hooks can *rotate 180 degrees*. You're allowed to be freaked out by that; I certainly am. Belemnites also bore arm hooks, though they differed in appearance from those of modern squid — and we don't know if any could rotate. Scientists think that hooks probably evolved independently in belemnites and in squid, perhaps from increasingly elaborate sucker rings.

As small as these structures are, it's possible to trace their evolution through rock because their hard material fossilizes rather well.[6] Squid hooks and sucker rings are made of the same tough material that coleoids use for their beaks: chitin. (Scientists and engineers have found an astonishing array of uses for squid chitin in recent years, from prosthetics for amputees to biothermoplastic for 3D printing.)

No suckers, rings, or hooks have been found on any fossil ammonoids or nautiloids, so these appendage accessories are considered one of the many exclusive coleoid inventions.[7] The ink sac is another, of course, and so is the breathtaking ability to change skin color, pattern, and texture. Modern nautiluses do not change their skin, and indeed the trick seems far less useful for an animal that keeps most of its body inside a shell.

Even the remarkable eyes of coleoids might have been part of the evolutionary package. Coleoid eyes are as complex as our own, with a lens for focusing light, a retina to detect it, and an iris to sharpen the image. Scientists have even outfitted cuttlefish with 3D glasses and found that their depth perception works like ours, comparing the information from left and right eyes.[8] Both coleoid and vertebrate vision evolved in swimming predators, so the common complexity isn't too surprising. But there are some striking differences. Our vertebrate retina has a blind spot where a bundle of nerves enters the eyeball before spreading out to connect to the front of each light receptor. By contrast, colcoid light receptors are innervated from behind, so there's no "hole" or blind spot. Structural differences like this show that the two groups converged on similar solutions through distinct evolutionary pathways.

Another significant difference is that fish went on to evolve color vision by increasing the variety of light-sensitive proteins in their eyes; coleoids never did and are probably color-blind. I say "probably" because the idea of color blindness in such colorful animals has flummoxed generations of scientists, and a few have suggested that modern squid and octopuses have the potential to exhibit unconventional kinds of color vision. Perhaps light-sensitive pigments distributed throughout their bodies could send color signals back to the brain.[9] Or maybe, by quickly changing the shape of their eyes, coleoids could scan through a series of wavelengths, comparing each new view to earlier views in order to see color over time.[10]

Color-blind or not, coleoids can definitely see something we humans are blind to: the polarization of light.[11]

Sunlight normally consists of waves vibrating in all directions. But when these waves are reflected off certain surfaces, like water, they get organized and arrive at the retina vibrating in only one direction. We call this "glare," and we don't like it, so we invented polarized sunglasses. Then, in an unrelated stroke of genius, we invented digital displays that produce polarized light, which is why your cell phone might look mysteriously blank if you check it while wearing sunglasses. That's pretty much all polarized sunglasses can do—block polarized light. Sadly, they can't help you decode the secret messages of cuttlefish, which have the ability to perform a sort of double-talk with their skin, maintaining color camouflage for the benefit of polarization-blind predators while flashing polarized displays to their fellow cuttlefish.

Such communication may seem surprisingly advanced for a creature without a backbone, but many species of cuttlefish and squid are quite social. Traveling in schools provides a pool of eligible partners when it comes time to mate and spawn, and offers protection from predators as well—another compensation, perhaps, for the loss of the protective shell. Certain species of squid may even hunt cooperatively.

It took time for all these amazing innovations to evolve. The first coleoid was a far cry from today's duplicitous cuttlefish or jail-breaking octopus. It probably looked just like an externally shelled cephalopod that woke up groggy one morning and accidentally put on its skin over its shell.

Jetting Out of the Shell

The internal shell of the most thoroughly studied early coleoid, *Hematites*, still had a hard tubular "living chamber" that contained most of the animal's meat. Its jet propulsion would have been of the less efficient shell-pumping type seen in modern *Nautilus*, wherein the animal forces water out of the siphon by pulling the body deeper into the chamber. Shortly, evolution would free the mantle from the strictures of the shell, opening the way to more efficient jet propulsion, supplemented with fins.

Hematites would have looked somewhat similar to the way we imagined *Sphooceras* a hundred million years ago in the Silurian, with a mantle covering the outside of the shell. It might even have had something else in common with *Sphooceras*: truncation. Many *Hematites* fossils are missing the first chambers of the phragmocone, which would have been formed in the egg and shortly after hatching, so a couple of scientists have theorized that this animal began its life with a partially external shell. After breaking off the "baby shell," the mantle would have grown around the remaining shell and secreted, not a mere cap as in *Sphooceras*, but a substantial shell guard, and the animal would have lived out the rest of its days with a fully internal shell.[12] Other *Hematites* fossils have been found with the embryonic shell intact, however, and the truncation theory remains unsubstantiated. What's certain is that the mantle always secreted a shell guard around the phragmocone tip.

The solid shell guard of *Hematites* would become de rigeur for coleoids over the next 200 million years, right up until the end-Cretaceous extinction. It was simply a more sophisticated version of the strategy used by *Endoceras* (that great big straight-shelled beast we met back in the early days of cephalopods), which deposited minerals into the chambers at the tip of its phragmocone to serve as a counterweight and keep the animal horizontal in the water.

Belemnites, the great coleoid success story of the Mesozoic, are never found without guards. Rather the reverse; the wonderfully fossilizable guard is often the only part of a belemnite to be uncovered. Enough complete fossils have surfaced, though, to show us evolution's belemnite breakthrough: reducing the tubular body chamber to a mere roof. Now the mantle, instead of wrapping around a shell, could hang from a rod (which bears the fancy name "proostracum"). The switch from tube to rod freed belemnites to develop the more efficient mantle-pumping form of jet propulsion, which works like blowing up a balloon and letting it fly.

You'll notice a pair of fins on that drawing of a belemnite, making it the first finned cephalopod to appear on our tour through

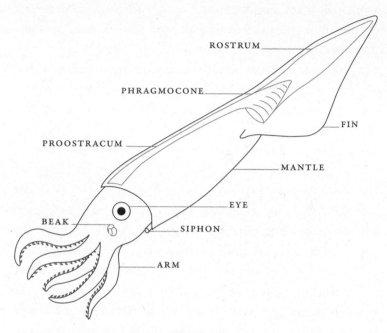

FIGURE 5.3 Although the fossils they most commonly left
behind look nothing at all like squid, when alive, belemnites were
very squiddy indeed. The notable differences are the ten equal arms
and the heavy "tail," which hides the shellguard.

C. A. Clark

the group's history. Why fins? Well, jet propulsion is inherently
less efficient than the paddling or undulating employed by fish,
marine reptiles, marine mammals, and basically every other swim-
mer ever, and cephalopods have converged on various techniques
to ease that inefficiency. Nautiluses evolved an alternative mode
of locomotion by wiggling the flesh of their siphons; coleoids
evolved fins. Some modern coleoids, like cuttlefish, now use fins
for nearly all their movement, reserving the jet for dire circumstances only.

However, although artists have been drawing fins on belemnites for nearly a century, their existence was purely speculative

for most of that time. Belemnite fins were first hypothesized by the Swiss cephalopod genius Adolf Naef, who was born in 1883. Like so many scientists of his generation, Naef published widely in multiple fields: systematics, paleontology, and developmental biology. But he always circled back to cephalopods. He was one of the first scientists to turn the common squid into a common laboratory animal, laying the groundwork for Hodgkin and Huxley's seminal work on giant axons. He cataloged the embryonic development of squid with stages that I still referenced routinely in graduate school in the 2000s.[13]

When it came to fossil cephalopods, Naef's careful observation of belemnite guards revealed paired grooves that would be the perfect place to attach fins. Ever since he published this idea in 1922, belemnite fins have enjoyed a healthy hypothetical existence. Further evidence mounted, including traces of blood vessels on some well-preserved guards. Finally, an extraordinary fossil discovery in 2016 brought fins into the realm of fact. A team of scientists including Klug, Fuchs, and Kruta found for the first time both fins and siphon in a fossil belemnite known as *Acanthoteuthis* (spiny squid) from Solnhofen, Germany.[14]

At first the fins were nothing more than faint suggestions in the rock. Shining ultraviolet light on the specimens and viewing them through special filters allowed scientists to see the full extent of the fins, preserved as ultraviolet-fluorescent phosphate. Ultraviolet light also exposed the siphon, along with two kinds of connecting tissue that hold the head and the siphon tightly to the mantle.

If you've ever dissected a squid or used one for fishing bait, you may have noticed the weak link between head and body, especially once the creature has been frozen and thawed. Pick it up by the head, and the body might fall right off—or the other way around. When the animal is alive, a muscular collar firms up that connection, while a special cartilage locks the siphon in place, "enhancing the effect of the water jet for fast swimming," according to Klug and colleagues.[15] Certainly the animal can jet more strongly if it's not worried about blowing its body apart.

In addition to fins and siphon, ultraviolet wavelengths illuminated a small, but telling fossil feature — statocysts, the cephalopod's equivalent of our inner ears. Statocysts are tiny chambers lined with hairs and filled with fluid, each containing a little rock called a *statolith*. When the squid accelerates, the statolith is pressed back in its "seat" like a person in a car. When the squid decelerates, the statolith rolls forward. Its movement pushes on the hairs, which send signals back to the brain. This is how the squid keeps its balance.

Modern cephalopod statocysts vary, depending on the animal's buoyancy and swimming ability. *Acanthoteuthis* statocysts look most similar to those of sleek muscular predators like Humboldt squid. The structure of the belemnite's fins, siphon, and collar all corroborate this comparison.

After warning, "It is impossible to confidently reconstruct the actual swimming speed of a prehistoric animal," Klug et al. take a stab at it anyway: "We thus speculate that belemnitids reached velocities between 0.3 and 0.5 m/s like, for example, today's *Todarodes* [Japanese flying squid] during migration."[16] That's about the speed of a dog paddle, enough to make headway against ocean currents, over long distances. And we don't know about belemnites, but modern squid can sprint. In short bursts, *Todarodes* reaches 36 feet per second (11 m/s). The human swimming record is 7.84 feet per second (2.39 m/s). Says Fuchs, "Olympic swimmers would need a lot more doping to compete with *Todarodes*, but possibly not to beat belemnites."[17]

Olympic-ready or not, though, belemnites had plenty of speed for catching fish.

My college biology professor exhibited childlike glee at every instance of an invertebrate preying on a vertebrate. I thought fondly of him when I read in 2019 that one such instance had been preserved for the ages. Four belemnites of the genus *Clarkiteuthis* were fossilized in the very act of chowing down on ancient fish. These belemnites got so distracted by their dinner that they sank into deep water, suffocated, and were buried. Tragic, but their fossils shed light on what was undoubtedly a successful fish-eating lifestyle for many other individuals.[18]

Even as the belemnite shell, with its solid guard and substantial phragmocone, proved its usefulness, other coeloids were busy taking shell reduction quite a few steps further.

Keep the Calcium

While continuing to calcify their shells and use them for buoyancy, two distinct groups of coeloids evolved away from the intricate three-part internal shell of belemnites. The first group was the cuttlefish, which utterly overhauled the chamber system, reshaped the shell, and retained only a trace of the shell guard. The second was the mysterious ram's horn squid, which even most cephalopod experts have never seen. These creatures kept the old style of chambers in their phragmocones, but the shape of the shell evolved all the way back to a coil.

A few minutes is sufficient to envision the millions of years it took for the cuttlebone to evolve. It's rather like what would happen to a belemnite's shell if you compressed the animal from its fins to its head. That slender rod of a proostracum disappears entirely, and the guard is spread out into a thin shield. Under this shield, the chambers of the phragmocone are both flattened and sheared, so they become diagonal layers. Pillars now hold these layers apart like a microscopic version of skyscraper scaffolding. The tube of the siphuncle turns into an abstraction that scientists call the *siphuncular zone*, which still extracts water from the spaces between the layers and allows gas to seep in and take its place.

A cuttlebone looks fragile, and it is. That's why it works so well as a calcium supplement for pet birds, as well as chinchillas, hermit crabs, and tortoises. A nautilus shell is full of calcium too, but the shell is so sturdy that even a very determined budgie would have a hard time extracting much of the necessary nutrient. The many thin walls of cuttlebones are far easier to crunch and grind. There's also not much cause to worry about the impact on wild cuttlefish of collecting their shells for this purpose. Unlike nautiluses, cuttlefish are abundant and short-lived, which means the ocean is just naturally full of cuttlefish dying, and their cuttlebones wash up abundantly on beaches. Cuttlebones don't have to look nice for a birdcage, so it doesn't matter if they've been broken or damaged.

The cuttlebone's fragility also limits the living animals, as it can't withstand the pressure of deep water. Cuttlebones of various species have been calculated to implode at depths of about 650–2000 feet (200–600 m). That's shallow by comparison with the depths of the sea, which number in the thousands of meters, so a great deal of ocean is off-limits to cuttlefish. (Still, they can get considerably deeper than most recreational scuba divers. My deepest dive was only 98 feet (30 m), and even there I began to feel the effects of nitrogen narcosis, a sort of drunkenness brought on by breathing ordinary air at high pressure. My buddy and I had brought a raw egg with us to crack at depth, and the way the water pressure held the egg together even without its shell was the coolest thing we'd ever seen.)

Deep water, however, is a welcoming home to ram's horn squid. Like cuttlebones, their uniquely coiled internal shells often wash ashore on beaches — I've found dozens in Australia — but these equally fragile-seeming shells are much more robust to pressure. Their small size no doubt helps save them from implosion; most ram's horn shells are no bigger than a quarter.

Ram's horn squid might be the most enigmatic of modern cephalopods. Their buoyant coil keeps them oriented head-downward,

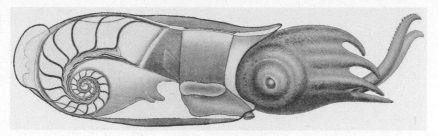

FIGURE 5.4 Ram's Horn Squid, showing
the shell in position in the animal.
Carl Chun, Die Cephalopoden, *1910*

and they have a large luminescent organ at the top of their body
that can glow for hours. They are yet another vertical migrant,
moving from depths of nearly 3,000 feet (1,000 m) during the day
to as shallow as my deepest dive during the night. Their eyes bulge
out to the sides, yet their entire head and arms can be retracted inside the mantle.

It might also be the loneliest of modern cephalopods, in an ab-
stract evolutionary sense. Just one species of ram's horn squid has been
described, and it's only distantly related to other squid and to cuttlefish. All of its closest relatives, a group called the spirulids, are fossils.

There are a *lot* of fossil spirulids, though, enough to show us the
probable evolution of the ram's horn coil. The critical missing link
is a species called *Spirulirostra,* which begins its life by growing a
coiled shell, only to straighten it out with maturity and eventually produce a guard that looks quite similar to a belemnite's.[19] It
seems likely that a genetic tweak here or there to control switches
caused the coiling of babyhood to continue through adulthood,
resulting in the modern ram's horn squid. Yes, it's evo-devo again!
The retention of childhood characteristics throughout life is actually a common theme in evolution — from salamander adults
keeping their baby gills to human adults retaining the head shape
of a juvenile chimp.

Like spirulids and cuttlefish, squid evolved from something like a belemnite, but they jetted a great deal farther down the road of shell reduction.

The Mighty Pen

Squid evolved away the shell guard. They evolved away the chambered phragmocone. And they evolved away the calcium. Their internal shell was pared down until it was no more than a simple pen, and this allowed the mantle to finally become all that it could be. It also presented paleontologists with a mysterious array of fossils that seem to be part octopus, part squid.

The pen, introduced back in figure 1.2, now bears closer examination. The first question is, can you really write with it?

I've been teaching squid dissections in primary and secondary schools since just after completing my own K12 education, and I've lost track of how many frozen bodies I've defrosted and how many cold mantles I've sliced open.[20] As a vegetarian with a deep affection for living squid, I sometimes regret this history, but I remind myself that the boxes of squid in the bait store would otherwise have been used to fish for halibut, cod, bass, even sharks. And it's pretty magical to see kids light up when they look through the squid's lens or find those balance-granting statocysts beside its brain . . . or draw with its pen.

Since the pen is fully encased by the flesh of the mantle, extracting it messes everything else up. I wait until we've already identified all the organs, then I show the students how to pull out the thin, transparent, stiff yet flexible vestige of a shell. As a grand finale to the dissection, they use the pen to puncture the ink sac — kept carefully intact until this point — and write their names. It works.

It's not as good as a quill though, and rather harder to get, so to my knowledge no one's ever used squid pens as more than momentary, whimsical writing instruments. The fancy scientific name for the pen refers instead to its swordlike shape: *gladius* (same root as "gladiator"). Despite having two different names that are most definitely not "shell," a squid's pen is constructed inside an organ called the *shell sac*. Understanding how this happens will help us see what evolution did to turn heavy armor into a hidden sword.

Baby cephalopods, even octopuses, all have a shell sac, which is comparable to the mantle folds that baby clams and snails use to make their shells. Both shell sacs and mantle folds depend on something called the *shell field*. In snails (or clams), the shell field is a temporary dent in the mantle, which deepens to a cleft. After making the cleft, the clam (or snail) covers it with a membrane. Once the membrane is in place, the cleft smooths back out to lie flat under the membrane, and the membrane turns into a shell. Strangely complicated? Yes, indeed! It seems the temporary cleft might be needed to provide support while the early shell is still delicate, but no one's completely sure.

Now, squid also have a shell field, recognizable to people who are habituated to looking at tiny bits of animal tissue. The squid's shell field, in fact, is the little embryonic cap that makes squid embryos look like nautilus embryos, and makes both look like monoplacophorans — those ancient flat-shelled mollusks. But instead of using its shell field to make a shell directly by the "cleft" method, a squid grows a ridge of tissue around the shell field, eventually covering the shell field completely. This now-enclosed space is the shell sac, and it's within this sac that the pen begins to form. So, despite *looking* like a baby nautilus or monoplacophoran, a modern coleoid never goes through even the briefest phase of externality: Its shell is internal from the get-go.[21]

Though smaller and simpler than an external shell, a pen still has some structure. The main "spine" of the pen is called a *rachis*, a name it shares with the central "spine" of a bird's feather (a meaningless but fun commonality between dinosaurs and cephalo-

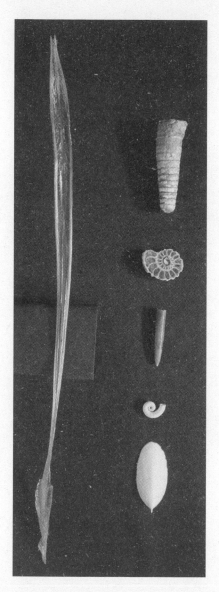

FIGURE 5.5

The story of the shell, as gleaned from specimens on my desk. *Left*: gladius of a Humboldt squid. *Right, from top to bottom*: fossil of a straight-shelled early cephalopod, fossil of an ammonoid, fossil of a belemnite, shell of modern *Spirula*, shell of modern cuttlefish.

Danna Staaf

pods). In some squid species, the pen rachis flares out into wings on either side, and the tip of the rachis grows into a thick cone. Some squid even have a further thickening of the pen beyond the cone, reminiscent of the shell guard of a belemnite. But no matter how thick and solid the tip of a squid's pen may be, it is never calcified. In this sense, it is never a true shell.

We know this from simply looking at pens, but we can confirm it by looking at the cells that *make* the pen. These cells lack the necessary machinery to calcify. In losing this functionality over the course of evolution, they paved the way for an entirely new kind of cephalopod to evolve.

Oxycone ammonoids might have skimmed through the water. Belemnites may have achieved substantial speed. But with the decalcified, semirigid, and semiflexible pen, modern coleoids could bend their bodies away from predators and make explosive escape jets. Dirk Fuchs and his colleague Yasuhiro Iba have named the pen "one of the key innovations responsible for the most powerful mode of jet propulsion among cephalopods."[22]

Such an innovation might have begun with the merest accident. A pen is nothing more complex than a decalcified shell, so one mutation of the genes that controlled calcification could be all it took. And fossil pens are pretty common, so this accident may have happened numerous times.

Because we associate modern pens with modern squid, there's a certain tendency to associate fossil pens with squid as well. A dizzying variety of early fossil coleoids with pens, or in some cases just the pens by themselves, have been described and given squiddy names (look for the "teuth" root): Plesioteuthidae, Teudopsidae, Trachyteuthidae. Further study, however, suggests that nearly all of them were closer to octopuses.[23]

As you can see from figure 5.6, some of the pen-bearing coleoids fossilized clearly enough for us to count arms, and they all had eight. A few even show suckers on their arms, and among modern

FIGURE 5.6 *Leptoteuthis gigas* is one of the mysterious
early gladius-bearing coleoids that seem to be most related
to octopuses despite a superficial similarity to squids.
Diego Sala

coleoids suckers offer a clear diagnosis: octopuses have *sessile* suckers sitting flat on the skin, while squid have *pedunculate* suckers attached to stalks like little mushrooms. The suckers that have been found in these pen-bearing coleoids? All sessile.

The final line of evidence comes from the pens themselves. A gladius does have some structure, and the rachis and wings and cones of all these early fossil pens match up better with an octopus pen than with a squid pen.

"But octopuses don't have pens," you may be thinking. Astute reader, you are correct. Their ancestors did, however, and the structure of the octopus pen can still be seen in one living species: the vampire squid, which really should be called a vampire octopus.

Vampire Feet

The story of octopuses and vampire squid is the ultimate reduction of the shell — from a gladius to mere vestiges, and in some species total loss. This group is known as the Vampyropoda, which is a weird word. It means "vampire feet," and no, that doesn't make any sense. It's simply the marriage of the two group names Octopoda and Vampyromorpha, conducted by scientists when they realized that octopuses and vampire squid are more closely related to each other than either is to any other cephalopod.

This took some realizing. Nothing will give you whiplash faster than trying to align a vampire squid with proper squid or proper octopuses. At first glance, its pen makes it seem like a squid. Then you notice that it has only eight arms like an octopus. Look a little closer, and you see two little filaments that might be highly reduced arms, reminiscent of a squid's tentacles. And then it turns out that those two filaments came from a different pair of arms than the pair that squid turned into tentacles — the filaments are the same arm pair that disappeared entirely from octopuses.

With genetic studies, scientists have finally agreed that the vampire squid is actually an octopus, an affiliation that renders its gladius highly relevant to the study of coleoid evolution. Close inspection reveals that the vampire gladius is entirely distinct from the squid gladius — so distinct that scientists believe they evolved from two separate decalcification events.

Ancestral vampyropods diverged from the ancestors of squid way back in the Triassic. Sometime in the Jurassic, as coleoids were reinventing themselves, the vampyropod lineage split — into vampire squid on one hand and octopuses proper on the other, both still bearing pens. Vampire squid kept theirs relatively unchanged. As with spirulids, only a single species of the once-prolific vampire lineage remains in our seas, and it seems to be something of a "living fossil."[24]

Meanwhile, octopuses proceeded apace along the path of shell reduction. Around the end of the Jurassic or beginning of the Cretaceous, octopuses had another evolutionary split. One lineage, the cirrate octopuses, kept the gladius as a single piece that eventually became shaped like a horseshoe. These are deep-sea octopuses with big floppy fins, the most noteworthy of which is named the Dumbo octopus (really). "Cirrate" refers to rows of short tendrils called *cirri* that grow along their arms. The cirri probably serve some feeding purpose, but observing a deep-sea octopus's dinner is difficult. It's almost as hard as trying to figure out how ammonoids ate.

Incirrate octopuses, or, as I like to think of them, real octopuses, have no cirri, no fins, and no pens. The pen of a vampyropod seems to be of primary use as a support and attachment point for fins, so without fins, there's no need for it. But before the octopus pen went away completely, it split in half. The paired remnants, called vestiges, are well known from the earliest fossil octopus, the still-finned *Palaeoctopus*, and their shape can be seen within that animal's outline in figure 2.4. Over time these vestiges were further reduced to two little rods called *stylets*, which are still found in some modern octopus species.[25]

Even octopuses without stylets almost certainly retain the molecular machinery necessary to build them. Once we've developed genetic lines, as in the vision of Eric Edsinger-Gonzales, we might be able to induce an ordinary California two-spot octopus to grow gladius vestiges, maybe even a full gladius, and just *maybe* even a calcified internal shell—working backward in time to replay the evolutionary changes that led cephalopods to where they are today.

Our knowledge of these evolutionary changes owes a tremendous debt to one particular location on the planet where the soft bodies of coleoids fossilized in abundance.[26] In 1883, when *Palaeoctopus* was first described, the rocks it came from were part of the Ottoman Empire.[27] In 1944, when the French paleontologist Jean Roger published "Le plus ancien Céphalopode Octopode fossil connu," the newly independent Lebanese government had just overturned French colonial rule.[28]

It's time to take a little detour into the intertwined history of humans and fossils.

Fossils in History:
From Fishing Fields to Buffalo Stones

People have noticed fossils of shelled cephalopods all over the world since ancient times, but soft-bodied fossils have been much harder to come by. They depend on the formation of *Lagerstätte* — German for "storage place," this term refers to a rock bed with phenomenal fossil preservation. (The drinkable kind of lager is an abbreviation of *Lagerbier*, "beer for storage.") Dozens have been found in many different countries from many different geologic times, preserving everything from dragonfly wings in Germany to those contentious *Nectocaris* fossils in Canada. The Cretaceous Lebanese *Lagerstätten* have been especially generous with coleoid fossils.

Herodotus wrote about them in 450 BCE, and the bishop of Palestine in the fourth century CE considered the fossils evidence of Noah's great flood. Centuries later, the visiting King Louis IX was given a stone in the shape of a "sea fish," and a few centuries after that, scientists began publishing in earnest about these remarkable rocks where octopuses with countable suckers lay alongside exquisitely preserved fish.

Throughout the early decades of Lebanon's independence, most people who lived near the fossiliferous quarries didn't spend much time there. "It's all stone and can't be cultivated, so they would help the foreigners for just small tips," says Roy Nohra, owner of the fossil museum Expo Hakel. "Many of the fossils you see in European museums were taken from here for almost nothing."[29]

In the 1970s, Nohra's father, Rizkallah, then a young boy, loved collecting fossils and dreamed of building a museum to house them. Civil war broke out in 1975 and dragged on for fifteen years, but still Rizkallah continued to collect. He even began restoring a small old house to create his museum. In 1991, the Lebanese parliament passed an amnesty law and the militias were dissolved; also

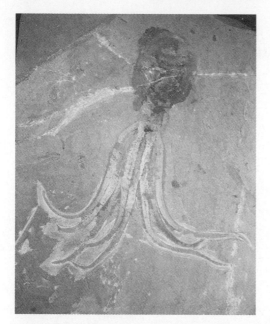

FIGURES 5.7A & 5.7B
Remarkable coleoid
fossils from the
Cretaceous rocks in
Hakel, Lebanon.
Roy Nohra, Expo Hakel

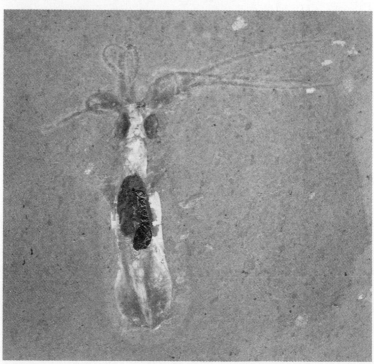

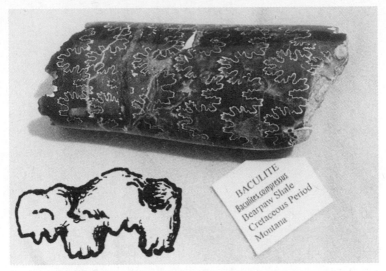

FIGURE 5.8 Ammonoid fossils like this one were collected by
Blackfeet and other Plains Indians as buffalo-calling stones. As the
fossil erodes, the isolated chambers form buffalo-like shapes.
Adrienne Mayor, Fossil Legends of the First Americans, figure 69

in 1991, in the little town of Hakel halfway between Beirut and
Tripoli, Rizkallah Nohra opened Expo Hakel.[30] "It's a small museum but still a very big step if you consider all the things happening at that time," says Nohra.

Since learning about the museum, I've added Hakel to the
short list of marvelous destinations I dream of visiting one day. I
wonder if anyone in the town is old enough to remember a visit
from Jean Roger. I hope he was kinder to the local people than
many fossil collectors and scientists have been throughout history.
Some ammonoids of North America, unfortunately, are among
the many fossils that have been appropriated from their rightful
owners without remuneration.

To Blackfoot and other Native peoples of the North American
plains, ammonoid fossils were known as buffalo-calling stones and
were believed to attract this valuable resource. At first I found the

connection mystifying, but a simple drawing made it clear. Those abundant straight-shelled heteromorphs called baculites often fossilized in fragments, their outlines defined by their fancy septa. A single chamber in isolation does look rather like a buffalo — especially if you've got buffalo on the brain.

Fossils like these were included in medicine bundles to bring good luck or healing to the individual who carried them and were passed down through generations. Sadly, many of these valuable heirlooms were stolen from prisoners of war after the Battle of the Little Bighorn in 1876.[31]

"The practice of taking valuable fossils from conquered lands or weaker people is not new, and the powerful emotions evoked by such acquisitions are not uniquely modern. Contention arises whenever rare and valuable geological objects come to light," wrote the Stanford scholar Adrienne Mayor, who focuses on ancient interpretations of fossils before the advent of the modern science of paleontology. "Large vertebrate fossils have long been tied to cultural identities and power inequalities. Those same links persist in modern-day fossil disputes in North America, in clashes between authorities and the people whose land contains geological objects of great scientific and monetary value."[32]

The most notorious case is the bitter conflict over the *T. Rex* known as Sue, excavated in 1992 and the object for the next five years of legal battles between the original landowner, the scientist who excavated her, the Sioux tribe, and the federal government. As Mayor points out, powerful entities have been stealing fossils from the less powerful at least since the ancient Greek city of Sparta (which you've probably heard of) snagged a mammoth skeleton from the ancient Greek town of Tegea (which, like me, you've probably never heard of). More recently, in the 1920s, the Gobi Desert saw a "bone rush" of North American paleontologists excavating dinosaur fossils and shipping them back home. Mayor

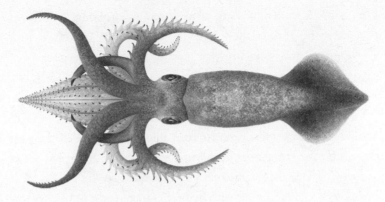

FIGURE 5.9 *Passaloteuthis* is one of many belemnites whose fossils have
been called "thunderstones" and used in folk cures.
Franz Anthony

describes the consequence succinctly: "The Chinese banned
Western paleontologists for the next 80 years, until 1986."[33]

Though less sought after than dinosaur bones, ancient straight-
shelled cephalopods are abundant in Chinese rocks, and the series
of septa between chambers is so reminiscent of the tiered eaves of
a pagoda that their name is *baota-shi*, pagoda stone. Like the ammonoids in Native medicine bundles, cephalopod fossils in China
have been used for medical treatment from ancient times to the
present day. Just on the other side of the Himalayas in India, certain coiled ammonoids fossils are named *saligrams*, symbols of the
god Vishnu, and are believed to offer spiritual rather than physical
healing.[34]

The English, too, attempted to use ammonoids to cure ailments
of both people and livestock, as did the Scottish and the Ger-
mans. Belemnites are also found abundantly throughout Europe
and were once used as medicine. They were called thunderbolts
or thunderstones, obviously having fallen to earth as "darts of
Heaven." One could activate their healing effects by either soaking
them in water or grinding them to dust. Thunderstone water was
supposed to cure distemper in horses; thunderstone dust could be

blown into a person's eyes to cure "soreness," which strikes me as perhaps a case of the remedy being worse than the disease.[35]

Ammonoids in England were called snakestones because people thought they looked like coiled snakes. Without heads, but never mind, legend can take care of little details like that, and artists could (and did) carve the heads back on. In this case, the legend was based on a real person, Saint Hilda of Northumbria from the 600s. The story goes that she planned to build a convent at Whitby, a place inconveniently full of snakes, so she cut off their heads with a whip and prayed the bodies into stone. Severe! A few hundred miles south of Whitby in Keynsham, the same praying-snakes-into-stone story is told about Saint Keyna, but Hilda's story is better known. She's the only saint whose name graces the scientific literature of ammonoids, in the genus *Hildoceras*.

This got me thinking: What if more fossils received the sort of nominative treatment Neil Shubin gave *Tiktaalik roseae*, the famous "fishapod" link between fish and four-legged beasts? Before choosing its scientific name, Shubin consulted Inuit elders from the Canadian territory where the fossils were found, and they suggested the Inuktitut word for a particular codlike fish. Dinosaur nomenclature, too, has been known to honor the language and culture of the people whose land housed the animals' bones. The quintessential flying dinosaur, *Quetzalcoatlus*, is among the most well known, its name a latinized version of the name of the Aztec god of wind. There are also *Zuniceratops* and *Anasazisaurus*, named for the Zuni and Anasazi peoples. There's even a shark, *Siksika*, which borrows its name from the Siksika Nation.[36] Among cephalopods, numerous ammonoid fossils like *Choctawites choctawensis* bear the names of Native peoples or places, but it's not clear whether any of this naming was preceded by respectful consultations like Shubin's. As more species are discovered, perhaps one day *Hildoceras* will be joined by scientific names recommended or informed by speakers of Blackfoot, Chinese, or Arabic.

6

Fall of the Empire

After reaching an acme of abundance in the Mesozoic, both ammonoids and belemnites were erased by another mass extinction—the same event that did away with every non-bird dinosaur. Scientists have known about that for generations, but only very recently has enough of the extinction story been pieced together that it's starting to make sense. Curiously, the stories of the two great cephalopod losses are turning out to be quite different.

The end-Cretaceous extinction of ammonoids was not a gradual case of turnover, of less fit species being replaced by those more fit, but one of abrupt doom brought about by an enormous meteor impact. Around the same time as the impact, and perhaps even related to it, the planet shuddered through another round of substantial volcanic activity. Both volcanoes and meteor drove major environmental changes that wiped out nearly all the ammonoids, probably because of their vulnerable babies. A few species did survive for a short time, and there's even a theory that ammonoids might still live among us as octopuses—though most scientists find little to support this idea.

Belemnites, on the other hand, make a more convincing case for continued existence. Their overall decline at the end of the Cretaceous seems to have been driven less by a sudden change in their environment than by competition with their own descendants. As one lineage of belemnites evolved into spirulids and, eventually, squid, other lineages were shouldered out of the seas.

Sudden Death

To understand what caused the end-Cretaceous mass extinction, the first challenge was to recognize that it actually happened. This recognition, instigated by the discovery of a meteor impact and supported by fine-scale fossil studies, constituted a major scientific shift from a paradigm of gradual decline to one of abrupt cataclysm.

For a long time, paleontologists viewed the tremendous turnover from a world of dinosaurs (and ammonoids and belemnites) to a world of mammals (and nautiluses and squid) as both slow and inevitable. This perspective of the extinction went along with the perspective, at the time widely accepted, that dinosaurs themselves were slow and archaic. They were obviously doomed.

Then in 1980, an iconoclastic father–son duo, Luis and Walter Alvarez, championed a new theory of extinction involving a catastrophic meteor strike.[1] As evidence for the meteor grew irrefutable, paleontologists remained puzzled over how such an event could have eradicated dinosaurs (etc.) while leaving mammals (etc.) untouched. Fine, there was an impact, they said, but it was just the last nail in the coffin. Dinosaurs were already deep in decline.

Ammonoids elicited a similar train of thought. For instance, the 1996 edition of the authoritative *Ammonoid Paleobiology* textbook tells us, "The ultimate extinction of the ammonoids was a continuous and long-lasting decline that can be traced over several

million years." This is followed by a decisive dismissal of the meteor strike: "There is no need to involve a major cosmic impact to
explain the final decline of ammonoids."[2]

But the acclaimed nautilus biologist and paleontologist Peter
Ward has argued passionately for a sudden demise of ammonoids.
In places where rock layers have been deposited continuously
from many years before the asteroid impact to many years after,
Ward and his collaborators have found that numerous ammonoid
species fossilized abundantly right up to the impact — and then
abruptly disappeared. Statistical analysis also showed that even
those ammonoid species that disappeared from the fossil record
well before the impact were likely to have actually survived to the
end of the Cretaceous. "Everything looks more gradual than it
was," explains Matthew Clapham. "If you happen to pick a given
rock, not every animal that lived then is fossilized in that rock.
So there are gaps within a species' life span. And if there are gaps
within, there must also be gaps at the ends. So not every species
appears to go straight to the boundary."[3]

After much scientific back-and-forth with various degrees of ci-
vility, the importance of the meteor impact in the end-Cretaceous
extinction has been widely accepted. In the latest edition of *Ammonoid Paleobiology* (2015), Landman and colleagues write, "It is
now generally accepted that the disappearance of ammonites . . .
was due to the asteroid impact. However, the exact killing mechanism is still unknown."[4]

So it was Mr. Asteroid in the Cretaceous Room with . . . the
candlestick? The rope? The revolver?

In Search of the Smoking Gun

The extraterrestrial shock that life received at the end of the Cretaceous was compounded by another round of extreme volcanism,
only a little less destructive than the one that ended the Permian.

Both the explosion from without and the explosions from within Earth set off a cascade of environmental changes that could have killed animals by heat, by cold, or by acid.

My daughter's preschool is embarking on a dinosaur theme as I write. Apparently they covered the Alvarez impact hypothesis at circle time; she came home with an enthusiastic description of the meteor "way up in outer space" that hit Earth and "tore off all the dinosaurs and plants." It's a compelling story, to be sure. I struggle to explain in four-year-old language that the meteor didn't physically knock everything off the planet. Instead, it started fires, changed the climate, damaged the air and the seas . . . I've lost her, and I realize I don't fully understand it myself.

There's some comfort in good company. "I realized after fifteen years working on mass extinctions I still don't know what causes them," said David Bond, a paleontologist from the University of Hull in England, at the beginning of a talk at the 2016 meeting of the Geological Society of America.[5] This opening statement was a bit glib, though, because he went on to explain quite convincingly what does cause nearly all mass extinctions: volcanism. And even though the end-Cretaceous extinction was anomalously triggered by a meteor impact, it also involved some pretty epic volcanoes.

Illustrations in dinosaur books tend to have lots of volcanoes in the background. Such prehistoric scenes sometimes even show a terrified *Tyrannosaurus* in futile flight, like a Pompeiian fleeing Vesuvius. However, volcanoes were not popping up all over Earth and erupting on a daily or even a weekly basis, as the picture books might lead us to believe. End-Cretaceous volcanism was concentrated in India, which at this point was still an island on its way toward a dramatic slow-motion collision with Eurasia. In fact, the volcanic flood basalts in India are very similar to the end-Permian paving of Siberia, 190,000 square miles (500,000 km^2; compared with the Siberian 772,000 square miles/2,000,000 km^2) and erupting over the course of perhaps thirty thousand years (to the

Siberian hundred thousand).[6] While less impressive, it's clearly volcanism on a grand scale and could have had similarly dramatic environmental impacts.

But . . . what about that meteor? Well, here's the wild part: *maybe the meteor caused the volcanoes.*[7] Some geologists suggest that the volcanic area in India was preloaded with magma, and the profound shock of a huge rock smashing into the planet brought this magma to the surface. It's hard to know for sure whether this connection is real, but the extraterrestrial rock in question would certainly have been capable of triggering global earthquakes. It left a dent more than 110 miles (177 km) wide in the Yucatán Peninsula in Mexico, now called the Chicxulub (pronounced "cheek-shuh-loob") crater. That's big enough for the entire big island of Hawaii to fit inside, with room to spare. Scientists calculate that the crater-producing meteor must have been at least 6.8 miles (11 km) in diameter, which means it could have comfortably contained Mount Everest (if the great peak had existed back then — since India was still an island, Everest had not yet protruded from the planet).

The impact had impacts both short- and long-term on every habitat — except the deep sea. Earth's surface was cooked for a few minutes by the reentry heat of particles that had been thrown out into space. Gases released into the atmosphere caused an "impact winter" in subsequent years (like the nuclear winter hypothesized by science fiction writers of the mid-1900s), as well as acid rain that dramatically changed the surface chemistry of the oceans.[8]

Obviously, if you happened to live at the end of the Cretaceous, you got a pretty raw deal. But the question that remains is how changes like these led to the specific changes we see in the fossil record. Why did some groups of animals disappear altogether, while others emerged relatively unscathed? One answer is simple chance. Lady Luck plays a huge role in extinction; each crisis is a new roll of the dice that sends some unfortunate species away from the table.

"Ammonoids almost bit it at every major extinction event," Kathleen Ritterbush points out, which suggests that they were vulnerable.[9] Though they finally did bite it at the end of the Cretaceous, it could have happened as easily at the end of the Permian or of the Triassic. The good luck that allowed them to survive as long as they did is perhaps more worthy of wonder than the bad luck that ended their days. Still, as Jocelyn Sessa notes, "It's our human nature that we are going to look at this [extinction] event and look for a reason, look for explanations."[10]

Ammonoid Survivors

It's not hard to imagine how a planet that got smashed with a mountain-sized meteor and spewed forth its molten innards could become a fairly hostile environment for life. The great puzzle is the specificity of the extinctions: why dinosaurs but not birds, ammonoids but not nautiloids? For cephalopods, the explanation may have to do with how far each species had spread itself around the globe. Those that covered more ground had more resilience.

The two groups of externally shelled cephalopods were certainly distinct from one another, but at first glance these differences should have favored the ammonoids. Ammonoids were diverse, abundant, complex. Nautiloids were simple, and while not exactly rare they were not common either. Yet nautiloids continued on their merry way all the way to the present day, while ammonoids perished at the end of the Cretaceous.

At least, that's what everyone thought, until some remarkable fossils cropped up near modern-day Maastricht in the Netherlands. As it turns out, ammonoids *didn't* go extinct at the end of the Cretaceous. Not all of them. Not immediately. A few species survived the apocalypse . . . for a little while.

Maastricht is a city familiar with both endings and beginnings, as the oldest Dutch city continuously inhabited since Roman

times and as the birthplace of the European Union. It has also lent
its name to the very last slice of Cretaceous time before the extinction; the Maastrichtian Age is named for the many fossils that have
been found near the city. The rock layers in this region, however, do
not end with the meteor strike; they preserve evidence of it and continue on into post-Cretaceous times, offering an excellent site for
comparing fossil assemblages just before and just after the cataclysm.

In 1996, just a few years after the European Union was signed
into being, John Jagt, a paleontologist at the Natuurhistorisch Museum Maastricht, discovered the narrative-challenging ammonoids that had survived the great extinction. Their fossils were
definitively preserved *above* the rock layer that marks the end of
the Cretaceous. They are three heteromorphs: one of the recurved
kind, *Hoploscaphites*, and two of the straight-shelled kind, *Baculites* and *Eubaculites*. Jagt went on to work with Neil Landman
and Peg Yacobucci (as well as scientists from Belgium, Poland, and
Russia — a cosmopolitan consortium) on the fossils' further study
and interpretation, and he was the one who kindly provided me
with a copy of their 2014 paper.[11]

It was a golden opportunity for these scientists to test hypotheses of why (nearly) all the ammonoids went extinct. What unique
features of the surviving species made them different from all the
ammonoids whose term ended with the Cretaceous? They tackled
the puzzle by spreading their fossil net out from Maastricht and
examining ammonoid geographic distributions around the world.
They looked at twenty-nine sites, from Turkmenistan to New Jer-
sey, Antarctica, and Egypt, using a computer program to slide
each site to its correct Cretaceous position. (As shown in figure
4.1, the continental arrangement was quite different back then. In
addition to India being an island, North and South America had
not yet collided.)

The scientists marked which of the twenty-nine sites had fossils
of which ammonoids, both survivors and not. Some were found at

only one site, others at many sites. By connecting the dots for each kind of ammonoid, they were able to estimate the size of its range. As it turned out, the ranges of the three ammonoid survivors were significantly larger than the ranges of the ammonoids that didn't make it across the Cretaceous boundary.

One kind of nautiloid that has been found on both sides of the end-Cretaceous boundary was analyzed as well, and it, too, was widely spread around the globe. The results imply that cephalopods with a broad range were more resistant to extinction than the others. From the "luck" perspective, it makes sense. Inhabiting a greater geographic range is akin to the proverbial multi-basket approach to egg storage.

"However, even the most broadly distributed ammonites eventually succumbed to extinction, whereas *Eutrephoceras* [the nautiloid survivor], with its smaller population sizes and larger embryonic shells, survived," Landman and his colleagues wrote in the sobering conclusion of their paper. "Evidently, a broad geographic distribution may have initially protected some ammonites from extinction, but it did not guarantee their long-term survival."[12]

Hmmm. Larger embryonic shells, you say . . .

Clues from Babies

We often consider survival to mean "not dying," and this definition works fine for the individual, but when it comes to species there is an even more important aspect of survival: procreation. Adults must avoid death long enough to make babies, and the babies must avoid death long enough to mature and make babies of their own. Ammonoids had been successful procreators for a long time, but it seems the very feature that made them so malleable to natural selection — their tiny abundant eggs — constituted a weak point in the generational chain when their environment went haywire.

Depending on its species, a baby ammonoid may have hatched in various places. One might have been cocooned with its siblings in a floating mass of jelly, from which it wriggled free into the water. Perhaps another was cushioned on seaweed or sand in the shallows near shore and began to exercise its minute jet as soon as it left the egg. A third could have spent its embryonic days nestled within its mother's shell, aerated by her gentle breath, only to be swept out to sea on that same breath after hatching.

No matter the species, though, it seems likely that all ammonoid babies grew up in the plankton. Too small to swim effectively against a current, they would have drifted hither and yon in a well-salted soup of other larvae, shrimp, worms, and more. This soup was both delicious and dangerous—with no nutritious yolk to fall back on, infant ammonoids needed to devour whatever they could catch, including their siblings. Meanwhile, many larger creatures (including unsentimental ammonoid parents) used nets and filters to consume vast quantities of plankton (including baby ammonoids).

When the asteroid struck, most plankton were devastated. One popular explanation for this die-off has been the idea that dust and gas thrown into the atmosphere obscured so much sunlight that planktonic algae died en masse. Today, as in the Cretaceous, these microscopic cells are factories that use the intangible rays of sunlight to manufacture the fundamental materials of life. As they bloom and are consumed by the little animals around them, which are food in turn for bigger beasts, algal cells fuel the food web all the way up to the biggest mosasaur (then) or whale (now). With algae killed off, starvation would have rippled throughout the ocean.

However, the latest calculations indicate that even the impact of a truly massive asteroid couldn't block out that much sunlight, so some further explanation is needed for the loss of life among plankton (and so many other life forms). Acid rain is currently the leading suspect.[13]

The acid could have come from both the meteor impact and volcanic eruptions, and when it dissolved in the surface waters of the ocean, the pH dropped precipitously. Acidification isn't always a big deal for every organism. Though low-pH water can deform or damage shells, studies on the impact of ocean acidification on modern mollusks have shown that creatures like baby clams can handle a slightly deformed shell. Baby ammonoids were probably not so robust.[14]

The survival of ammonoid infants depended on their phragmocone, the tiny gas-filled shell that kept them afloat. About the size of a rice grain, this minuscule ammonitella had much thinner walls than an adult ammonoid's shell. A thin shell can be grown quickly, but it also leaves the animal vulnerable in a suddenly acidic environment. Damage the shell and the young ammonoid would lose its buoyancy, sinking away from the only environment it was adapted to live in.

Although also born with small phragmocones, nautiloid babies would have been far less affected by the acidification of surface waters. An order of magnitude larger than baby ammonoids and much slower to develop, they could have simply waited out the unfortunate fallout of Chicxulub. A modern nautilus embryo, Yacobucci explains, "grows for more than a year inside the egg; before it hatches it's so big the shell is poking out of the egg!"[15] If ancient nautiloids had a similarly long incubation time, it might have brought them through the worst of the crisis, while the abundant yolk and large size of hatchlings allowed them to roam and scavenge food even in the absence of plankton.

"People have suggested that because the ocean was an inhospitable environment, because the nautilids had this storehouse they could rely on to grow versus the ammonites that would have to be relying on food from the environment, maybe that's why the nautilids were able to survive," says Sessa. "It is an appealing explanation."[16]

Unfortunately, Earth's history isn't a game of Clue, and there's no envelope we can open to check our theory of the ammonoids' extinction by Mr. Asteroid in the Cretaceous Room with Acidification. Instead of a definitive yea or nay, Earth offers us as many hints as we are willing to look for. There will always be more fossils to uncover, more scanning techniques or statistical analyses to apply, and with each new discovery we can test our theories anew. Some theories, like the once far-fetched idea of a tremendous asteroid ending the Cretaceous, have marshaled so much supporting evidence that they are virtually undisputed.

Others, like the idea that octopuses are naked ammonoids, well ... they're still a little far-fetched.

Wishful Thinking

We've probably all wished at some point in our lives that dinosaurs had survived for us to marvel at. Although there's unlikely to be a real instantiation of *Jurassic Park* anytime soon, we can take some solace in the modern world's bizarre avian pageantry, remembering that ostriches and penguins, finches and vultures are all direct dinosaur descendants.

A few scientists, perhaps with a little too much hope in their hearts, have trotted out the idea that ammonoids could still be with us in the same way. Could today's octopuses, from the aquarium giant that first elicited my awe to the little two-spot that lived in my bedroom, be "nude ammonoids" that managed to survive the end-Cretaceous extinction? The suggestion was raised as early as 1865. Adolf Naef, that prolific cephalopod polymath, proposed it in writing in 1922.[17] An Israeli paleontologist, Zeem Lewy, resurrected the idea in 1996.[18] The remainder of the paleontological community remains thoroughly skeptical.

The idea is inspired by the curious egg case of one of the most unusual of modern octopuses, the argonaut, which lives throughout the world's warm tropical and subtropical oceans. We met argonaut males earlier, as the shrunken bearers of those tremendously enlarged sperm-transferring arms called hectocotyli. There had been some scientific confusion over that arm, and it was initially given the name "hectocotylus" when it was thought to be a parasitic worm. Female argonauts have also caused their own share of scientific naming confusion. The female argonaut, in fact, was the original "nautilus."

"Nautilus" means "sailor," and the female of this particular octopus species was so named because two of her arms have great sail-like flaps. Not completely irrationally, the people who first observed and described the creature thought she might use those arms to catch the wind, sailing across the water's surface like a little boat.

Later, when the name "nautilus" was also given to the somewhat larger cephalopods with hard shells found throughout the Indo-Pacific, the two animals were differentiated by adjectives: the "paper nautilus" was the sail-armed octopus, and the "pearly nautilus" or "chambered nautilus" was the "proper" nautilus with the mother-of-pearl shell. More recently, to avoid confusion, the term "paper nautilus" has been left mostly by the wayside in favor of this small octopus's scientific name: *Argonauta*.[19]

The "paper" part of the old name referred to the delicate nature of the female argonaut's egg case. Though its shape resembles that of an ammonoid shell, it's far more delicate—the argonaut's body shows through the translucent walls. And it's no true shell. Instead of secreting their egg case with a shell sac, as nautiluses secrete their shells and even squid secrete their pens, female argonauts "Spiderman it out of their arms." (Truly this is one of the great verbings of our time; thanks are due to Kathleen Ritterbush for producing it.) Those two wide flaps on her arms, instead of catching wind, produce the material of her egg case, which she uses throughout her life as both mobile home and nursery. It's unattached to her body,

FIGURE 6.1 The modern female argonaut uses her arms
to spin herself a shell, which serves as an egg case
and helps maintain her own buoyancy.
Julian Finn, Museums Victoria

so she can crawl freely in and out of it—unlike modern nautiluses, which die if extracted forcibly from their shells.

Despite this freedom and despite the total absence of siphuncle or septa, the female argonaut has converged with her extremely distant relatives on the use of a coiled shell for buoyancy. She eschews any attempt to pump water out of her shell, instead simply swimming to the sea surface and positioning the shell to acquire an air bubble. With the bubble in place, she swims vigorously downward until she has reached a depth of neutral buoyancy, then goes about her business.[20]

The similarities between argonauts and ammonoids are remarkable: the form of the shell, its use in buoyancy, and even the possibility that some ammonoids might have laid eggs in their shells as modern argonauts do.

With a good dose of imagination, Lewy expanded on these ideas.[21] After laying eggs in their shells, he wrote, female ammonoids would have died, thoughtfully providing a corpse for their offspring to devour upon hatching. (Feeding one's offspring with one's own body is hardly exceptional in the animal kingdom; though we mammals manage to survive the transfer of nutrients, other species routinely make the ultimate sacrifice.) He proposed that argonauts evolved from these ammonoids in the Cretaceous, lost their own shells, and then began to lay their eggs in the empty shells of other ammonoids. Then, he suggested, they started to use their arms to mend these old shells or add onto them. The eventual extinction of the ammonoids obligated argonauts to make their egg cases from scratch, based on the ammonoid blueprint.

Gerd Westermann, the creator of the hydrodynamic triangle of ammonoid shapes, promptly published a paper with a colleague dismantling Lewy's argument.[22] They pointed out that there's no evidence of any ammonoid shell ever being modified by an argonaut, and anyway, modern argonauts construct their egg cases with a different kind of material than that used to make ammonoid shells (calcite instead of aragonite). Another problem is the depth in time of the octopus lineage. Octopus ancestors roamed the oceans of the Jurassic and maybe even the Triassic, well before the appearance of late Cretaceous ammonoids. These early octopuses are clearly not argonauts, which indicates that argonauts are a more recent derived form rather than the rootstock of all octopuses.

What to make of the visual similarity between argonaut shells and ammonoid shells? They're probably just evolutionary answers to the same question: how to swim efficiently underwater.[23] No paleontologist that I spoke to found anything substantial in

FIGURE 6.2 The late Cretaceous *Palaeoctopus* was a prominent
member of the long octopus lineage.
Franz Anthony

Lewy's arguments. Still, his revival of Naef's idea that octopuses
evolved from ammonoids reminds me of a modern revival of another of Naef's ideas: that squid evolved from belemnites.

In the case of octopuses and ammonoids, several generations of
scientists have considered, evaluated, and rejected the idea. In the
case of squid and belemnites, the idea was initially discredited by
the prevailing view of belemnites as an evolutionary dead end for
coleoids. However, contemporary scientists are now finding more
and more evidence to confirm Naef's original vision.

Hiding in the Deep

For lunch it is possible to eat a modern-day dinosaur; for dinner,
a modern-day belemnite. Or you could switch it up if you prefer
calamari for lunch and chicken for dinner.

Of course, just as a great many dinosaur species did go extinct,
so did many belemnites — not only at the end-Cretaceous event,

but throughout the period leading up to it. Unlike dinosaurs and ammonoids, belemnites really did experience a long-term decline well in advance of the meteor impact, perhaps due to competition with early squid. Modern ten-armed cephalopods (the group that includes squid, spirulids, and cuttlefish) were once thought to have arisen from a coleoid lineage independent of belemnites, but recent work by Dirk Fuchs and Sasha Arkhipkin, among others, has brought science back around to Naef's view, in which squid could be considered the greatest belemnite success.

I first heard Arkhipkin speak at a scientific conference when I was a giddy undergraduate. I'd scored some funding from my college to attend the triennial meeting of the Cephalopod International Advisory Council (CIAC) in 2003 in Thailand, and everything about the experience was a surreal thrill, from bunking with other penny-pinching students in the Phuket Marine Biological Center to hunting octopus with a local fisher who knew how to extract a writhing mass of arms from a barren scrumble of rocks. Though I was awestruck by all the great luminaries of cephalopod science gathered there, I had little idea what any of their talks were actually about.

Seven years later, during which time Arkhipkin had done a term as president of CIAC and I had wrestled a PhD from mounds of data, I met him in person at the 5th International Symposium on Pacific Squid in La Paz, Mexico. Having transitioned from giddiness through desperation to resignation over the course of my doctoral study, I felt a resurgent thrill of interest in brand-new science as Arkhipkin described his latest big idea.

"This is the rostrum," he said, scribbling a diagram on a scrap of green paper, and I stuck with him through the thick Russian accent and the equally thick Mexican heat because the picture he drew was legitimately fascinating. It was the tip of the pen of a modern oceanic squid, like the Humboldt squid I'd been so immersed in for the past six years—and it was chambered like the phragmocone of an extinct belemnite.

The drawing on the paper was not actually meant to be a Humboldt squid's pen because this species doesn't swim around the Falkland Islands, where Arkhipkin lives and works. Its local equivalent is the Argentine short-fin squid, a sizable fishery resource that Arkhipkin is tasked with managing, as senior fisheries scientist of the Falkland Islands government. Part of managing a fishery is understanding how old the animals are and how fast they grow, and one way to figure that out for squid is to count the layers of chitin in the pen, like tree rings. As he was performing this routine task, Arkhipkin suddenly encountered "these very strange structures almost like septa."

He laughs now as he describes his attempts to diagnose the structures. "I contacted paleontologists to send them the picture of that rostrum, and they immediately said it was from belemnite, and I said, 'It's not belemnite, it's a modern squid,' and after that it was silence, because it was against the theories, that belemnites were outcasts, that they didn't produce any offspring for modern squid."[24]

Argentine squid weren't the only ones with "septa" in their pens. In the early twentieth century, Naef had seen them in some species of Mediterranean squid and had even suggested their affinity with the chambers of belemnites. But no one had pursued this idea further, until Arkhipkin began looking at other species. He found the same chambers in numerous other open-ocean and deep-sea squid. With Dirk Fuchs and one of their Russian colleagues, he published a paper in 2012 using the chambered rostrum to suggest that some belemnites moved into deep water to survive the end of the Cretaceous.[25]

This theory of deep-water belemnites hinges on the possibility that a mutation might have caused some animals to lose the ability to pump water out of their phragmocones. These nonbuoyant belemnites wouldn't have been able to compete with their fully functional relatives in the same environment. But what if they colonized a new environment, one from which gas-filled phragmocones were excluded? There would be no competition.

Such an environment is the deep sea, where creatures with gas-filled shells dare not venture for risk of implosion. In the late Cretaceous, the deep sea would have been void of other cephalopods, as this was long before modern depth-lovers like giant squid evolved. Most cephalopods still had shells that kept them in relatively shallow water, so the deep sea was a place where the mutant sinking belemnites could find refuge. Once they were ensconced there, evolution would quickly decalcify and reduce a shell no longer serving its buoyant purpose. In the deep sea, Arkhipkin hypothesizes, "maybe they survived the bolide impact and live now as squid."[26]

The logic is plausible and the chambers in the rostra are highly suggestive, but when considering whether members of one group evolved into another, scientists always hope to find a "missing link"—a fossil form intermediate between the two groups. In 2013, Dirk Fuchs found just such a fossil in North Pacific Cretaceous rocks. The newly christened *Longibelus* seems to be part belemnite, part spirulid.[27]

Cretaceous Earth was a warm greenhouse, with the continents well enough separated to humidify the dry deserts of Triassic Pangaea, and with not an iceberg or glacier to be seen. But the North Pacific was an anomaly. Though global warming was the rule, between Japan and California shifting ocean currents cooled the water instead. This didn't sit well with the local belemnites, which had all adapted to life in warm water, so by the mid-Cretaceous they had moved out.

Now, plenty of cold-adapted belemnites lived in the Arctic and could have spread south into the freshly hospitable Pacific— but before they could do so, the sea level sank and a land bridge emerged to close the Bering Strait. As dinosaurs began to migrate between North America and Asia, the North Pacific was cut off from the Arctic, leaving it permanently empty of belemnites. It became an ideal "nursery ground" for modern cephalopods. *Longibelus*, possessing a belemnite-like shell but no shell guard, showed up in the North Pacific just as belemnites evacuated the premises. After identifying it there, Fuchs and his colleagues reexamined fossils from around the globe and discovered that *Longibelus* had probably evolved back in the early Cretaceous, in the part of the Tethys Sea that was gradually shrinking into the Mediterranean. From there it had spread east into the Indian Ocean, and thence into the newly vacated North Pacific, where it flourished. By the end of the Cretaceous, *Longibelus* occupied nearly every ocean.

The "missing link" *Longibelus* likely gave rise to the first ancestor of all modern ten-armed cephalopods that we know from the fossil record, which is named *Naefia* in fitting tribute to Adolf. *Naefia's* internal shell is simple enough for the animal to be considered a true spirulid. Both *Naefia* and *Longibelus* seem to have preferred somewhat deeper water than belemnites, though they still had functional phragmocones that limited their depth. It's possible that some of their descendants were the ones to lose buoyancy and sink even deeper, thereby surviving the mass extinction that ended both the dinosaurs' rule on land and the ancient cephalopods' rule in the sea.

With a fiery combination of volcanoes and meteor strike, we must bid goodbye forever to stegasaurians and ceratopsians, ammonoids and belemnites. Time to greet to a new world inherited by their descendants: birds and squid.

7

Reinvasion

The past 60 million years have seen the cephalopod survivors of the end-Cretaceous catastrophe morph into the aquarium exhibits and calamari rings of today. During this time, they adapted to yet another new wave of predators and to a cool new world: the Cenozoic, the era of "modern life."

Nautiloids in the modern era went through a process of population reduction and relocation. Their initially global, shallow-water distribution shrank to a few local, deep-water populations. Meanwhile, coleoids carried on with various styles of shell reduction and sought safety from predators in the evolution of sharp eyesight and complex behavior, alongside their ink clouds and camouflage. The expansion of their behavioral repertoire may even have constituted a critical advantage that allowed them to move back into shallow water from their deep-sea refuges.

Many coleoids also moved farther along a fishlike evolutionary trajectory, and their story must be considered in light of the continued evolution of vertebrates — not only fish, but also whales and even humans. Swimming marine vertebrates have acted as antagonists toward cephalopods for as long as both groups have been around. We saw this antagonism first from the early Devonian fish, then from the Mesozoic combination of fish and reptiles, and now in the Cenozoic we'll see it from fish and mammals. Peering at the Cenozoic era through different lenses will allow us to put together a complete picture. First is the lens of climate change, including the onset of glaciation and the concurrent movement of cephalopods back into coastal waters. Next, the evolution of whales and the radiation of modern fish emerge as strong selective pressures toward habitat shifts and final shell reduction in coleoids. The total loss of the phragmocone in modern squid brings us to the development of ammonia for buoyancy, which had the curious side effect of erasing squid from the fossil record. Nautiluses, meanwhile, have somehow managed to retain their external shells: Are they doomed relics of the past or harbingers of evolution to come?

Finally, we'll bring ourselves up to date with the human-dominated Anthropocene, settling into the contemporary shape of the world as we look around us at the results of the past 500 million years.

Return from the Deep

After an initial warm spell following the end-Cretaceous extinction, continental drift and enthusiastic plant growth cooled the planet and kick-started a global ocean circulation pattern that would maintain itself up to the present day. Perhaps facilitated by these changes, deep-water coleoids reinvaded shallow water, developing and honing their eyesight, camouflage, and behavior along the way.

The "warm spell" was really more of a "hot spot." About 10 million years after the meteor impact, global temperatures rose enough to be given a name — the "Paleocene-Eocene Thermal Maximum" — and an intense amount of study by modern scientists, including Jocelyn Sessa, who worked with Neil Landman on the methane seep ammonoids. This thermal maximum is considered the best model within our planet's history for the sort of global warming that we're creating today, as it was caused by a similarly excessive release of carbon dioxide.[1] It happened over a much longer period of time, though, and the planetary life that experienced it was much different than today's flora and fauna, because nearly all of it had evolved in an already warm environment.

Many animals responded to the heat spike by migrating. On land, they went north or south, toward the poles and away from the equator. Some marine animals did the same, but they also had the option of moving deeper. Few creatures were truly cold-adapted, so it didn't matter that there was nowhere truly cold to live.

Then Earth's climate began a lengthy switch from warm and melty to cool and icy, possibly thanks to a single energetic plant named *Azolla*. This fast-growing aquatic fern could have grown so quickly and abundantly that it slurped huge quantities of carbon dioxide out of the air — the same gas that had been belched out by the end-Cretaceous volcanoes in India, creating a greenhouse Earth.[2] Normally such carbon drawdown would simply get recycled back into the air, as bacteria or animals consumed the plant material and breathed the carbon out. But with the right conditions, masses of *Azolla* could have sunk quickly to the ocean bottom and been buried before decomposing, constituting a one-way transfer of carbon from the atmosphere into the earth. As its blanket of carbon dioxide thinned, Earth rediscovered the joys of glaciers and ice caps.[3]

Continental drift also played a significant part in this transition. South America began to head toward North America while Australia moved toward Asia, both of them leaving Antarctica behind at the South Pole. The ocean was now free to slosh around this southernmost continent in a continuous circle, creating a circum-Antarctic current that further isolated Antarctica. In the maps of Pangaea and its subsequent breakup (figures 4.1A and B), you can see Antarctica begin its lonely polar tenure. (Don't feel too sad. In another hundred million years or so, future Antarctica is almost certain to get cozy again with either Australia or South America. Or both.)

Once Antarctica chilled enough for the seawater around it to start freezing, a strange chemical fact came into play: there is no such thing as saltwater ice. When seawater freezes, it leaves its salt behind. The water around the forming ice thus becomes even saltier — and saltier water can stay liquid at colder temperatures. So whenever a bit of seawater freezes, it creates two things: freshwater ice and extremely salty, extremely cold liquid water. The saltier and colder water becomes, the heavier it gets, so this frigid saltwater sinks to the bottom of the sea, following the dips and ridges of the seafloor to spread across ocean basins. That's why the modern deep sea is so cold.

This was the beginning of a global "ocean conveyor belt" that still flows today, carrying profound implications for everything from weather patterns to fishing grounds. Deep cold water not only flows along the seafloor, but is drawn back up to the surface in "upwelling regions," where algae feast on the rich nutrients it bears. These regions tend to occur along coastlines, like the California Current, where I learned to scuba dive amid vigorously growing kelp, prolific anemone gardens, bright orange garibaldi, and the ever-inquisitive harbor seals.

Initially, fish may have dominated these productive coastal waters, while cephalopods were relegated to oceanic and deep water. We can't be certain, of course, but the idea that at least some coleoids spent a period of their evolutionary history in the deep sea is supported by their lack of a phragmocone (which would implode at depth) and their survival of the end-Cretaceous extinction (when surface waters were acidified). But Neale Monks has pointed out an interesting inconsistency: in the deep sea, what use are good eyesight and visual camouflage — the hallmarks of modern coleoids?

Perhaps these traits didn't evolve until coleoids reinvaded shallow water. As Andrew Packard notes, "Such a course of evolution is reminiscent of the mammals that, for a long period of their early history, occupied a peripheral place in the world dominated by archosaurs, and whose special senses, especially vision, are thought to have reached a high degree of development in association with crepuscular habits. What allowed cephalopods to reinvade shallow water—whether it was the disappearance of the fishlike reptiles, or the increasing differentiation of coastal habitats creating new niches, or some new behavioral advantage acquired during their period in the wilderness—we can only guess."[4]

Behavior is certainly a key aspect of the convergence between cephalopods and fish. Both developed large brains relative to their forebears, and both are frequent subjects of fond behavioral anecdotes from hobby and professional aquarists alike. My first pet octopus, Serendipity, used to play tug-of-war with me by gripping her food just enough to offer some resistance but not enough to pull it from my hand—not until she'd had her fun. Later, long after Serendipity had expired, I kept a pufferfish named Agnes in the same aquarium in college. If I was slow to feed her in the morning she would rise to the surface and splash with increasing vigor until breakfast arrived. She also once spat a well-aimed and substantial stream of water at a visitor, who ever after afforded her tank a wide berth.

Scientists have corroborated such tales with empirical studies of tool use in octopuses. In the wild, these animals collect and arrange rocks to modify their dens, likely improving defensibility against predators. Octopuses also use their water jets to move sand, blow away waste, or repel unwanted visitors—reminiscent of pufferfish Agnes's attack on my friend. In one study, researchers gave octopuses an empty bottle and observed that, after bringing it to their mouths to answer the initial "Is this food?" query, the animals began to treat it like a toy. Several octopuses used their water jet to send the bottle circling around the tank, over and over, which reminded the researchers of a human bouncing a ball.[5] There's a delight that comes from recognizing play behavior in an animal so different from us, and it also illustrates the evolutionary importance of exploration and experimentation. These abilities are especially valuable in the constantly changing coastal environment, whether coral reef or rocky tidepool.

So far we've just discussed octopuses, of course, and people often wonder if squid are equally intelligent. Unfortunately, squid intelligence is a difficult study, because these animals are accustomed to swimming freely in the open ocean and don't take well to captivity. Octopuses, which will make themselves at home in soda bottles as readily as rock dens, are more amenable to laboratory life. Hungry and curious, they'll readily unscrew jars and solve mazes to find a morsel of crabmeat. Squid aren't built to handle this kind of task, but they have their own adaptations. As gregarious animals that clearly communicate with one another, squid are probably the first cephalopods in which we could look for evidence of language.

Several scientists in 2003 described a set of communicative skin displays in the Caribbean reef squid that they dubbed "squiddish." They found that a single squid could even hold two "conversations" at the same time; the classic example was a male squid displaying courtship toward the female on his left and aggression toward the male on his right. Adapting to a daytime existence in the well-lit shallow waters of coral reefs has afforded these squid plenty of opportunity to evolve nuanced visual communication.[6]

In general, colonizing coastal waters was a risky move for coleoids, as these places tend to be highly desirable real estate. Plenty of nutrients means plenty of competition and predation. However, hugging the rocks and beaches would have allowed coleoids to escape the notice of one particularly big and terrifying new kind of predator: whales.

Enter the Whales

The Cenozoic has been called the Age of Mammals, and with good reason. Mammals flourished all over, from the first primates and rodents on land to the first bats in the air, and, of course, the first whales and dolphins in the sea. The importance of these warm-blooded predators in the oceanic ecosystem can hardly be overstated, and they've certainly had a powerful influence on cephalopods. Coevolution of marine mammals with coleoids in particular may have led to echolocation in whales and final shell loss in squid.

According to Neale Monks, "While the mass extinctions at the end of the Cretaceous were bad news for the ammonites, the nautiluses sailed through them seemingly unharmed. Indeed, there is quite a flurry of new nautilus types immediately after the Cretaceous–Tertiary boundary, suggesting that to some degree at least nautiluses were able to take advantage of the niches left empty by the extinction of the ammonites."[7] The situation is comparable to the rise of mammals that followed the extinction of dinosaurs. Primates might never have been able to evolve and diversify if the dinosaurs hadn't been knocked out, and likewise nautiloids might never have filled the seas again if ammonoids hadn't been extinguished. As it was, nautiloids thrived through the warm years of the early Cenozoic.

Monks points out that even though ammonoids and belemnites went extinct at the end of the Cretaceous (or very shortly thereafter), their ecological "niches" were quickly refilled. The early flowering of nautiloids included some with increasingly complex sutures and fancy ornaments, like the ammonoids of yore. A striking example of an ammonoid-like nautiloid bears the magnificent name *Aturia ziczac* in reference to the intricate shape of its septa. And many early coleoids, especially spirulids, emulated the belemnite lifestyle by growing substantial internal phragmocones.

Meanwhile, the first whale ancestors were trotting around land on four legs, perhaps foraging for food in rivers and lakes. Eventually evolution led their descendants to swim in and drink the freshwater; they probably ate fish and crabs and maybe clams, but they wouldn't have encountered a single cephalopod. Never have the head-footed mollusks ventured out of the brine. Over time, continued evolution granted whales the ability to drink saltwater, and as they roamed through the expansive seas they discovered a bonanza of edible tentacles.[8] Being extremely large—the early whale *Basilosaurus* grew to more than 50 feet long, comparable in size to the largest ichthyosaurs or modern sperm whales—with large toothy maws, these early cetaceans wouldn't have had much trouble cracking a nautiloid shell or catching up with a fast-swimming coleoid.

But early cetaceans most likely relied on vision to hunt, and so could have been foiled by vertically migrating prey hiding beyond the reach of sunlight. Both coleoids and nautiloids likely engaged in this tactic, keeping to deep dark water during the day and rising to the surface only at night. It's been hypothesized that the echolocation of whales and dolphins evolved specifically to target cephalopods, allowing marine mammals to find and catch their prey even on the darkest nights.

As defined in the Pixar/Disney movie *Finding Dory*, echolocation is "the world's most powerful pair of glasses." It's a good analogy for us humans, who tend to be obsessed with vision, but in actuality echolocation operates more like an incredibly powerful hearing aid. The whale clicks and listens for the echoes of the click as they bounce back from surrounding surfaces. Human hearing is pitiful compared with that of most other animals—I have a hard time distinguishing between the whining of my cats and my children, let alone figuring out which direction the whining comes from—so it's almost impossible to understand how a whale could hear *so clearly* that it can create a soundscape of its environment, identifying rocks, fellow whales, and a calamari dinner. But that's what echolocation can do.

Poor cephalopods. There's no way to camouflage yourself when you're being targeted by *sound*. Or is there?

Echolocation works best when the echoes can bounce off hard surfaces. The softer the target, the more difficult it is to locate—in the extreme case, a jellyfish would be virtually invisible (or more properly, inaudible). And we must remember that, by the Cenozoic, many coleoids had already evolved away their hard shells. Back in the Cretaceous we met *Palaeoctopus*, with its gladius reduced to vestiges, and *Naefia*, the early spirulid whose shell-less squid descendants may have literally descended into the deep. Such modern-style squid and octopuses would have been soft enough to challenge an echolocating cetacean. They might even have wholly escaped the notice of early whales, while these predators concentrated on spirulids and nautilids.

"A hollow shell will return a very obvious echo, making the bearers of such shells easy targets," Neale Monks points out.[9] Although the beginning of the Cenozoic saw a diversity of cephalopods with hollow shells, this diversity was drastically reduced in concert with the spread of cetaceans. Today we have nearly ninety cetacean species, from killer whales to bottlenose dolphins, and not a single ammonoid-like nautiloid or any belemnite-like coleoids. The sole remaining spirulid has evolved its own suite of unusual features and, along with the remaining nautiloids, is geographically quite limited. Perhaps whales shifted the cephalopod balance in favor of soft-bodied squid and octopuses.[10]

While abandoning their shell, however, squid did not have to give up on buoyancy. They evolved a new technique, an approach to buoyancy that is invisible to echolocation and that, incidentally, may have rendered squid nearly unfossilizable.

Invisible Evolution

Paleontologists have inherited plenty of fossil cuttlefish and spirulids from Cenozoic times. There are even a few octopuses. But not a single squid. The best they can get is a few statoliths, those tiny inner-ear stones that help squid balance during swimming. And maybe a squid beak here, a hook there. "But no shell remains or soft parts," says the coleoid expert Dirk Fuchs. "Compared to the cuttlefish and the spirulids, almost nothing."[11]

Yet we are quite certain that squid existed throughout the Cenozoic. Where are their remains? It's always possible that we've simply been squid-unlucky. Maybe the fossils are still waiting somewhere to be found. But a far more satisfying answer has just been revealed, through a combination of rock chemistry and decay experiments: as it turns out, squid evolved ammonia, and ammonia kills fossils.

Jakob Vinther describes decay experiments as "one of the most atrocious things I have ever done."[12] Though the name is fairly innocuous, the reality of decay experimentation isn't — it consists of carefully observing and documenting the fate of a dead body. Vinther first tried this while working on fossilized ink for his PhD (the work that would lead to the discovery of melanin in dinosaur feathers). The incredible preservation of fossilized ink sacs fascinated him, and with decay experiments he thought he might gain some insight into how the preservation could occur. "I failed miserably," he says. "One of the reasons why was because when the squids decayed, the ink just splayed all over the place."

The disgusting disintegration of dead squid struck Vinther as a distinct contrast to the excellent preservation of so many octopus fossils. The critical component of high-quality octopus fossils seems to be phosphate, a chemical that occurs naturally in seawater, and in living tissue as a building block of DNA and of energy-carrying molecules. When conditions are right, the soft tissues of a dead animal can be quickly replaced with the mineral calcium phosphate. Such phosphatic fossilization sometimes preserves detail down to the level of single cells. Maybe there was a reason octopuses phosphatized and squid never did.

Vinther knew from research done by his PhD supervisor that phosphatization requires an acidic environment in and around the body. Eventually he found a student, Thomas Clements, who was willing to measure the pH of decaying cephalopods.[13] Sure enough, the pH of a dead octopus quickly dropped low enough to enter the *phosphatization window*. (As a measure of acidity, pH can seem counterintuitive. The lower the number, the more acidic the measurement.) But the pH of dead squid stayed far too high for phosphatization until "they had completely turned into disgusting mush."

The culprit that causes high pH in squid seems to be ammonia, which most modern squid use for buoyancy.[14]

You may recall that table salt is sodium chloride, and as the usual source of table salt is the ocean, we can conclude that most of the ocean's salt is also sodium chloride. But other salts can be made with other materials; ammonium chloride is one. Ammonia is a tiny bit lighter than sodium, so saltwater made with ammonium chloride is less dense than ordinary seawater — in other words, it's buoyant. Because the difference is so slight, huge amounts of ammonium chloride are needed to create significant buoyancy, and indeed some ammoniacal squid (as they are called) devote more than half of their mass to ammonia-saltwater.

Stick with me for just one more chemical fact: ammonia is a base, the opposite of an acid. An un-acid, if you will. We know that an acidic environment is needed for phosphatization. Ammoniacal squids are by their very nature basic. Calcium phosphate doesn't have a chance of replacing their tissues, and instead they turn to unfossilizable goo.[15]

Squid lost the phragmocone, the buoyant part of their shells, perhaps because of evolutionary pressure from echolocating whales. Or perhaps it made them more effective swimmers, better able to compete with fish. Or perhaps it was just an accident. Whatever the reason, buoyancy itself remained a desirable feature, so in the absence of gas an alternative buoyancy mechanism evolved: one involving ammonia.

As a side effect with no implications whatever for the squid, but with profoundly frustrating implications for the modern paleontologist, squid lost the ability to fossilize. Says Vinther, "They basically evolved themselves out of the fossil record."[16]

Living Fossil?

"A really good question is why *Nautilus* still exists," says Jakob Vinther bluntly. "I have no answer to that. That thing, when you look at it, it's swimming around banging into stuff; it hardly has an ability to see what's around it — I have no idea how that thing can still exist."[17]

Nautiloids, as we've seen, have been around for a long time. A really, really, really long time. And their modern representatives don't seem to have changed all that much, earning them the frequently tossed-about title of "living fossil." With their slow, awkward swimming and limited eyesight, nautiluses seem hopelessly out of date. Yet they are not dumb relics of the past, any more than large dinosaurs were doomed to be superseded by mammals. (Although this idea has been discredited and may be unfamiliar to young dinosaur lovers, the myth survives when we speak dismissively of old technology, or out-of-touch bosses, as "dinosaurs.") In fact, the more we learn about nautiluses, the more complex and intriguing they become.

Nautiloids seem to be the great endurers of the cephalopod lineage. They glided through each mass extinction that nearly wiped out ammonoids, including the final annihilation. The key to their ongoing survival might be the relatively large amount of energy they put into each nautiloid baby. A hefty store of yolk may have freed nautiloid hatchlings from a dependence on plankton, allowing them to wait out the bad times. This same ability to wait it out may have brought them success throughout the Cenozoic. Although their experiments with ammonoid-style complexity didn't pan out, they still seem to be holding their own—even diversifying anew.

One clear advantage of increased parental investment in each offspring is the opportunity to teach skills and pass on knowledge. Most of the animals we think of as intelligent take full advantage of this opportunity. Elephants and dolphins, for example, birth just one calf at a time and spend years guiding its development. Guppies and flies, which have yet to demonstrate complex behaviors such as tool use or mirror recognition, spawn dozens to hundreds of babies at once and offer no parental care at all.

So it's rather odd that nautiloids, which have never impressed anyone with their intelligence, carefully produce only a few babies, while coleoids, the indisputable brains of the invertebrate world, leave their multitudinous offspring to fend for themselves. Even a brooding coleoid mother like an argonaut, who watches and helps her eggs to hatch, has no interaction with her hatchlings. She doesn't feed them or guard them; she doesn't teach them to hunt or help them to hide. This lack of interaction isn't unusual among invertebrates, but it is unusual for a big-brained animal with the capacity to learn. The amazing behavioral repertoire of adult cephalopods—their tool use and play, their mixed mating messages (remember the cross-dressing sneaker males?) —becomes that much more amazing when you realize that they develop it in complete independence, with no parental instruction, over less than a year. Modern nautiloids, which have kept the safety of a shell, are typically considered less intelligent, since they have simpler brains and eyes.

However, this assumption has rarely been tested. Although nautilus brains lack the lobes that coleoid cephalopods use to form memories, in 2008 a couple of scientists decided to find out whether nautiluses could remember that a flashing light was associated with food.[18] Twelve nautiluses were trained à la Pavlov with a flash of blue light and a tealike infusion of frozen tilapia head, which caused them to breathe hard and wave their tentacles in excitement. At time intervals from minutes to days after their training, the animals were shown a blue flash without any tilapia tea—and to everyone's surprise, these "primitive" cephalopods showed the same signs of excitement, proving that they had formed both short-term and long-term memories.

Nautilus eyes have been similarly sidelined. Certainly coleoid eyes are remarkable, possessing lenses like those of vertebrates and producing high-resolution images. But in the absence of lenses, nautilus eyes have evolved impressive image formation with a mere pinhole. If you have personal experience with myopia, you probably know that squinting—taking in light through a smaller opening—can sharpen the image. Nautiluses basically squint a lot, all the time. Theirs is not as good as coleoid vision, to be sure, but nautiluses rely more on smell than on sight anyway. Unfortunately, we humans are not very good smellers. Visual intelligence makes sense to us. Olfactory intelligence is a serious puzzle.

The modern nautilus may seem unchanged, but there are obvious and then there are subtle changes. The most obvious changes are gross anatomical ones, the changes that we see easily by comparing fossils with their living relatives. That's how we track the dramatic shell reduction of coleoids. But nautiluses may have undergone similarly dramatic changes in ways that aren't as easy to track. Their olfactory abilities may be a recent innovation. We don't even know what sensory abilities lurk within their plethora of tentacles. And like every other creature on Earth, nautiluses continue to evolve.

During the early Cenozoic years, nautiloids were found all around the globe. Sometime after the onset of cold water and whales, perhaps driven in part by these challenges, nautiluses disappeared from most places, to be found only in Australia. It might have seemed that this was the last refuge of a group nearing extinction. In support of this idea, when modern nautiluses were first described by scientists, they were thought to comprise but a single species.

Eventually more dead nautiluses and empty shells made their way from the tropical Indo-Pacific to Europe, where they were categorized into four or five species—two of which were known exclusively from shells. In the 1980s, Peter Ward and his colleagues found both species alive in the wild, and one of them, *Nautilus scrobilatus*, revealed a tremendous secret. Its shell was covered with a fuzzy layer of organic material called a *periostracum*, which is familiar from certain sea snails but had never before been seen on a cephalopod. The soft insides of this species were also strikingly different from those of other nautiluses, and eventually, in 1997, *scrobilatus* was given a whole new genus—*Allonautilus*, or "other nautilus." This was a hint that nautiluses might not be such an evolutionary dead end after all.

Further work revealed that nautilus populations of all species fall into one of two groups: either they are found in the seas around Australia, or they are found much farther out to sea, near isolated islands like Palau and Fiji. This distribution indicates that nautiluses have begun to spread again beyond their Australian constriction. Could nautiluses be on the verge of a new radiation? Genetics indicate that *Allonautilus* is a young genus, recently birthed from within the *Nautilus* species group, which suggests there's a lot of work that evolution could do with this lineage.[19] Who knows if new nautilus species may even reinvent ammonoid-style sutures or ornament?

Unfortunately, humans may not give evolution a chance. The mass extinction we are in the midst of creating poses a different

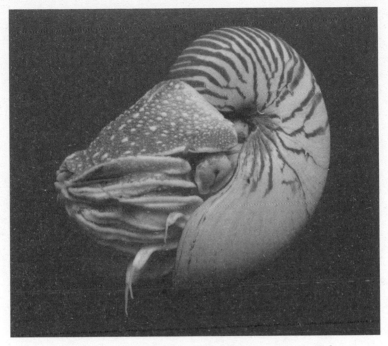

FIGURE 7.1 A living nautilus swims in the water near Palau,
displaying its many specialized tentacles, fleshy hood,
pinhole eye, and countershading stripes.
Wikimedia Commons user Manuae

kind of threat than all the mass extinctions nautiluses have survived thus far.

Closing Time

Over the course of the Cenozoic, the continents moved into the
positions they occupy today—though of course only temporarily,
as they are still and forever on the move. As various bits of ocean
were separated by land, Panthalassa finished fragmenting into the
"seven seas." These changes helped to create the evolutionary patterns we see throughout the world, both aquatic and terrestrial,

and they also might have created the current ice age we find ourselves in . . . an ice age that human industry is altering at unparal-
leled speed, with tragic results for our fellow Earthlings.

A scant dozen million years after its wild volcanism subsided,
the Indian subcontinent crunched up into Asia. The Himalayas
began to rise, and rise. Another collision followed to the west,
as Africa brought the Arabian Peninsula to Europe, closing up
Tethys and birthing the Mediterranean. Finally, the Panama isth-
mus connected the Americas and separated the Pacific and Atlantic Oceans. This strengthened coastal currents on either side
— the Humboldt Current in the Pacific and the Gulf Stream in
the Atlantic — both of which worked to further cool the planet.
The Gulf Stream in particular donated abundant tropical mois-
ture to the air above it, which was then carried up to the Arctic,
where it could fall as snow and build the polar snowpack.

The current ice age began 2.5 million years ago, marking the
onset of the Pleistocene, an epoch famous for glacial cycles — and
for the evolution of *Homo sapiens*. Ice coverage varied throughout
the epoch in perhaps a dozen alternating glacial and interglacial
periods. Despite the popular use of "ice age" to refer to the time of
woolly mammoths and migrating humans, when glaciers covered
both New York and Old York, that time is more accurately known
as the last "glacial period." It ended about 15,000 years ago when
permanent ice retreated to the poles and high places and the interglacial Holocene began. That's our contemporary epoch, although
scientists are considering that perhaps the Holocene ended and
the Anthropocene began when humans became the most influen-
tial species on the planet.

From our first appearance in Africa we moved around the
world, often settling near the ocean, swimming and fishing and
rowing and sailing. Just like the first animals in the Cambrian Explosion, we're actively changing our environment, rendering many
ecological niches inviable and opening a few new ones. There's

a huge difference, though, in terms of the *rate* and the *scale* of change. While microbes drastically changed the composition of Earth's atmosphere a couple billion years ago, it took them hundreds of millions of years to do so. Humans have been around for only a hundred thousand years, and we've been seriously altering the atmosphere for only a paltry few hundred years.

As for scale, we're pretty ambitious. Sponges may have pumped nutrients into the seafloor and opened it up for more animals to live, and fish may have eaten a few groups out of existence, but humans have devastated entire ecosystems and driven hundreds if not thousands of species extinct.[20] We've warmed the planet and accelerated the rise of the seas.

Certainly these things have happened in the past. Sometimes studying the geologic history of the planet can give us the wrong impression, making us think that global climate change is no big deal. But as Jocelyn Sessa says of the Paleocene-Eocene Thermal Maximum, "It is the best analogy we have. But as far as we can tell, that was happening on the order of thousands of years, and what we're doing is orders of magnitude faster than that."[21]

One of the results of change happening more quickly is that species have less time to adapt to the change. Temperature in the Paleocene rose slowly enough that animals exposed to higher temperatures had already evolved in a fairly warm climate. By contrast, every creature that faces global warming today has evolved in a cold climate. "It'll never be a perfect analogy," says Sessa. "I do worry more about modern mollusks because they evolved in an ocean that never saw more than 32 degrees [Celsius], even in the tropics. When we've done studies on modern mollusks, they're typically pretty unhappy at high temperatures."

In the oceans, temperature may not even be the main challenge life faces. "We often think about the temperature question and we don't think as much about the acidification question," Sessa notes. "Or we think about them separately, but they're going to be happening at the same time." As we've seen, acidification from the ocean's drawdown of carbon dioxide affects shell-building animals, making it harder for them to create shells in the first place and dissolving the ones that are made. It also affects chemical signals, leading to that chilling "death by celibacy" hypothesis.

Another source of aquatic doom goes hand in hand with rising heat and acid. Think back to all those mass extinctions, where global climate change driven by volcanism or extraterrestrial impact wreaked havoc on life. Temperature changed, yes. Acidification may have occurred (it's hard to tell from rocks). But the key killer in many of these cases was *lack of oxygen*. Warm water holds less oxygen than cold water, and organisms can suffocate because of it. In 2013, the scientists of the International Programme on the State of the Ocean, in conjunction with the International Union for Conservation of Nature (IUCN), coined a term for the threat the oceans face: a "deadly trio" of warming, acidification, and lack of oxygen.[22]

The interconnected nature of the planet is often discussed in ecological terms, and the mutual dependence of life on other life is indisputable. But even the chemical and physical systems of the world are deeply entwined with each other, and with us living components. Changes wrought by humanity are now being felt on every level, from the deepest ocean trench to the most remote mountain glacier. After much informal bandying about of the word "Anthropocene," a group of scientists in 2016 officially recommended using this name for a new geologic epoch. The International Commission on Stratigraphy is deliberating as I write.

That's how things stand now, 500 million years after the first fossil cephalopod and 200,000 years after the first fossil human. It's time to meet all the living, jetting, inking, thinking cephalopods we share the planet with.

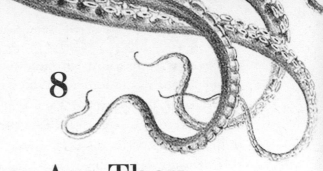

8

Where Are They Now?

The study of modern cephalopods falls under the purview of marine biology, a field aspired to by many a dewy-eyed dolphin lover but eventually pursued in earnest only by those with a high tolerance for dirt — not just actual dirt, but fish guts too — and frustration.

The frustrations of marine biology are surprisingly similar to those of paleontology. The depths of the sea can be nearly as mysterious and inaccessible as the depths of time. It simply isn't possible for us terrestrial animals to fully explore the ocean, so we're forced to take samples. Likewise, we haven't yet figured out how to travel backward in time. So the fossil record serves as a sample of the diversity and abundance of past life. Unfortunately, the sampling technique in both cases tends to kill the specimen.

Reflecting on the challenges of gathering paleontological data, Kenneth De Baets points out, "It's a bit similar, actually, if you are a marine biologist. Mostly it's just based on indirect evidence: stomach contents, or you catch *some* of them, but you cannot follow the entire population."[1] Both fields of study also suffer from a great mushing together of evidence. Animals that lived thousands of years apart are often mixed in one fossil bed, and animals living a few miles apart in the sea are often mixed in one trawl net.

Our technology for exploring and sampling the deep sea is becoming ever more sophisticated, but still the most basic of techniques — drag a net through the water and see what it catches — is a staple of oceanographic research. Learning from such a trawl is rather like mounting an archaeological expedition; fragments of life must be identified and fit together. Not all animals are caught equally by the net; not all are treated the same once caught. The most delicate, gelatinous creatures simply disintegrate, often leaving no trace of their existence — a modern echo of their near-total absence from the fossil record.

And yet paleontologists can work wonders with careful study of the fossils they do find. Marine biologists, too, can give us remarkable views into the lives and habits of the biggest, smallest, weirdest, and scariest modern cephalopods.

How Many, How Much?

Compared with the wildly diverse speciation of past cephalopods (ahem, ammonoids, I'm looking at you), the modern members of this group are relatively modest. By number of species at least, snails are the true molluscan masters of the world. However, certain individual species of cephalopods reach staggering abundances.

Taken all together, mollusks comprise tens of thousands of named species and probably an order of magnitude more as yet unnamed.[2] Within the mollusks, the heavyweights in terms of sheer diversity are definitely the gastropods, those "stomach-footed" true snails, which are estimated to comprise more than a hundred thousand living species. There are probably quite a few more minuscule snail species hiding out in sands that have never been sifted and studied. None of the other groups of mollusks come close to this, but the bivalves — modern clams and cousins — still number more than ten thousand species. At only around eight hundred living species, cephalopods are among the less species-rich mollusks. (Let's not make them feel bad by comparing them with the insectoid wealth of arthropods. Nobody needs that.)

Nor are cephalopods as diversely situated as snails. In addition to the terrestrial garden snails we're all familiar with, there are snail representatives in deserts and rainforests, in saltwater and in freshwater. Cephalopods, meanwhile, are limited to the ocean.

You may have heard rumors of a Pacific Northwest tree octopus that has adapted to life in the damp forests of Oregon and Washington. Alas, this species is merely a charming hoax begun in the early days of the Internet — and kept up to date through the succeeding decades.[3] I found the website through my dial-up modem connection in high school, and though I was a pretty naive kid I was also a passionate cephalopod nerd, well aware of the complete absence of terrestrial octopuses. Riding the rare pleasure of being in on a joke, I bought one of the site's T-shirts: it had a sucker-covered arm twisted in the shape of an awareness ribbon and the words "Save the Pacific Northwest Tree Octopus." I still treasure it.

Although real cephalopods can't live on land, even in the moistest of forests, octopuses can survive for a time out of water by holding a mantleful of seawater to breathe. This is how they accomplish their great escapes, like Inky's dash to freedom from the New Zealand aquarium — though the ability to venture across dry land has been useful to octopuses since long before the invention of aquariums. Many wild octopuses live in shallow tidepools, wetting their gills in no more than a puddle before sallying forth in pursuit of a scuttling crab. In one curious reversal at a California marine reserve, an octopus crawled out of the water, approached a group of human bystanders, and left a dead crab at their feet before returning to the sea. "What a friendly dude," commented the videographer who managed to record the event.[4]

Let any cephalopod remain too long away from water, however, and it faces death by suffocation and desiccation. Even freshwater is hostile to cephalopods. The closest evolution has come to crossing this barrier is the adaptation of certain cuttlefish to mating and spawning in the brackish water where rivers mix with the sea. They tolerate salinities much lower than that of the open ocean, but still brinier than tap water—and they've never been known to swim upstream. Alas, our world knows no river squid, no lake octopus.

But when it comes to considering the planetary habitat, these aren't significant limitations. Freshwater covers a measly 0.15 percent of the earth's surface;[5] land is somewhat more substantial at 29 percent but still doesn't hold a candle to the oceanic environment: 71 percent of surface area, vastly more by volume. The ocean is a layered environment, one habitat atop another atop another, and different species adapt to life in different layers. Some cephalopods live at the very rim of the sea in tidepools; others lurk in the abyssal depths. None have yet been found at the bottom of the Mariana Trench, but there's no a priori reason they couldn't live there. Certainly cephalopods have been found on the dark seafloor, participating in unconventional vent communities, like the methane seep ammonites we met in the Cretaceous.

Between shallow tidepools and deep seeps lies the disorienting open blue with nothing to break it—no surface, no rock, no seaweed. It's hard for us to even imagine this as a valid living space. What would you do? What would you eat? But their ability to thrive in this midwater blue may be one of the cephalopods' greatest successes, illustrating that what this group lacks in species abundance it makes up in numerical abundance. The tremendous biomass of midwater squid allows countless dolphins, whales, walruses, and seals to build their bulk out of a squid-heavy diet, and despite the constant predation these squid continue to proliferate and flourish.

Then there are giant squid. Humans haven't seen many, ever, and we had never seen a live one in its own habitat until 2004. So you might start to think they're pretty rare. But then you look in a sperm whale's stomach, and you see giant squid beaks. Hundreds of giant squid beaks. About the size of your palm, they may not seem impressive on their own, but contrast them with the corn-kernel-sized beaks of ordinary market squid that you might eat with bread crumbs and tartar sauce. The hand-sized beaks found in every sperm whale stomach ever opened definitely came from giant squid.

If each sperm whale in the ocean were to eat one giant squid a week, more than 18 million giant squid would be eaten by sperm whales every year. But consumption might be far more frequent than that. A single whale might even eat multiple giants a day. Scientists have estimated that throughout the world's oceans, sperm whales might be eating 3.6 million giant squid, every single day.[6] One cannot help but imagine a thick layer of whale chow in the deep ocean that simply writhes with tentacles.

Of course, now that we're talking about giant squid, we can no longer avoid the question of size.

How Big?

Lord Alfred Tennyson's glorious poem is one of many worthy contributions to the cephalopodan sea monster myth:

> Below the thunders of the upper deep,
> Far, far beneath in the abysmal sea,
> His ancient, dreamless, uninvaded sleep
> The Kraken sleepeth: faintest sunlights flee . . .

Many tales of imaginary beasts may well have been inspired by real giant squid; even the myth of the sea serpent might have started with a waving arm or a striking tentacle. A squidlike terror that can tackle an entire ship and swallow its sailors like so much popcorn is too good a story to let go, although science forces us to accept that no squid really grows that big.[7]

Let's have a look, then, at some actual monsters.

The giant Pacific octopus, which mesmerized my childhood self along with innumerable other aquarium visitors, fuels its growth with food from the cold, nutrient-rich waters off Alaska, Canada, and the rest of the US Pacific Northwest. The largest members of this species are comparable in weight to an adult human, with a maximum of 156 pounds, and possess an incredible arm span measuring more than 10 feet from tip to tip.

At least one other octopus species, the inaccurately named seven-arm octopus, may grow as large as or even somewhat larger than the Pacific giant. (When not in use, the male's hectocotylus hides in a pouch, leaving him to go about his business with only seven arms. A relative of the argonaut, he is also much smaller than his mate, so it is the female seven-arms that are the size record holders.) As big as these octopuses get, the contenders for the title of largest cephalopod are indubitably squid.

For a long time, there wasn't much competition — there was simply the giant squid, *Architeuthis*. But then in 2007 a new species seized the world's attention: the colossal squid, *Mesonychoteuthis*, the one with rotating hooks instead of suckers.

We know even less about colossal squid than we do about giant squid, so it's hard to say how big they really get. What's more, length is a troublesome measurement. Even the arm span of an octopus is questionable — measuring an octopus from arm tip to arm tip is like measuring a human from outstretched fingers to toes, instead of the more common technique of measuring from heel to crown.

But at least octopus arms are relatively constant in length. Squid tentacles are not. Intrinsically elastic, often scrunched into little pockets but able to shoot out at great speeds to great lengths, should they really contribute to the animal's length at all? Of

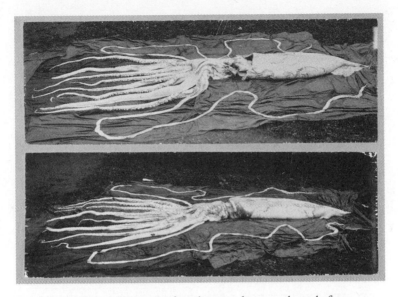

FIGURE 8.1 Giant squid are known almost exclusively from
dead specimens washed ashore, like this individual
photographed at Hevnefjord, Norway, in 1896.
NTU Museum of Natural History
and Archaeology

course not, but here are the numbers for total tentacle-inclusive
length anyway: 32 feet for colossal squid, 42 feet for giant squid.
Giant squid are the clear winners.

However, scientists use neither tentacles nor arms when comparing cephalopod lengths; their preferred measurement is mantle length. For the biggest squid, this is substantial enough! The
mantles of both giants and colossals have been reliably measured
at more than 8 feet; the tallest human could lie inside such a mantle and be completely hidden (though it would smell pretty bad).
By this standard, the two species are tied for the size prize. One
more measurement is available, though, and it turns out that colossal squid can actually outweigh giant squid by as much as a few

hundred pounds. The "biggest squid" title can thus be assigned to either species, depending on whether you deem weight or length to be the critical factor.

Enormous squid are spectacular, it's true. But tiny squid are adorable — and I say cuteness trumps spectacle. Furthermore, tiny squid have faced and solved a truly curious conundrum: How does jet propulsion work when you're the size of a rice grain?

The problem arises because water does not behave the same at every scale. You may have noticed an ant, for example, become helplessly stuck in a drop of water, while a water strider can dance across a pond. Any tiny creature encounters water as a thick, viscous medium; walking on top of it makes more sense than trying to swim through it. Yet squid are constrained by their family history; their bodies carry the legacy of jet propulsion. How to adapt?

To figure it out, I spent a few years on intimate terms with the world's tiniest squid: babies, of course, about the size of short-grain rice. Spending hours watching them through acrylic walls and more hours watching videos of them on my lab's old CRT screen, I decided that "jetting" wasn't an appropriate word for their minute movement. I decided to call it "squidging." As I amassed measurements of mantle expansion and contraction, I eventually concluded that their wee mantles have evolved a shape more like bells than torpedoes, which allows them to swim with the gentle strokes of a jellyfish instead of the powerful pumps of a large squid. It's not terribly efficient, but it works.[8] As the animals mature the bells grow into torpedoes, and the squidges grow into the huge exhalations of the fastest squid on the planet — because the very smallest baby squid happen to grow up to be the very largest adults. Giant and colossal squid, as well as the Humboldt squid I studied, all spawn rice-grain-sized babies.

The smallest *adult* squid, by contrast, have evolved in a different direction. We're talking here about adult squid less than an inch long. They're called pygmy squid, and yes, their jet propulsion is

FIGURE 8.2 Pygmy squid have been successfully
cultured in captivity and displayed at the Monterey Bay
Aquarium's *Tentacles* exhibit since 2014.
© *Monterey Bay Aquarium*

inefficient their whole lives long. Their solution: don't bother to
swim. Pygmy squid spend most of their lives stuck to blades of seaweed with an absolutely fantastic little glue gland. Exactly how the
glue works is still under scientific scrutiny, but two equally wondrous possibilities have been proposed: The glue gland could be a
"duo-gland" with one type of cell secreting a sticky mucus for attachment and another type of cell secreting an acid to dissolve the
mucus when release is needed. Or the two cell types could secrete
different kinds of mucus that work like two-part epoxy, effecting
an attachment only after they've been mixed together. In this scenario, the squid would probably just squirm itself loose whenever
necessary.[9]

Despite the name, pygmy squid are more closely related to
cuttlefish than to true squid, as their internal shells are more like

cuttlebones than pens. The group of cephalopods called sepioids includes bottletail and bobtail squid in addition to the well-known cuttlefish and the lesser-known pygmy squid. As hinted at by the pygmy's glue gland, sepioids house some of the flat-out weirdest members of the modern cephalopod family.

How Weird?

As with so many other features, various cephalopod species have converged with various fish species on features of extreme weirdness. Multipurpose mucus—check. Luminescent lures and glow-in-the-dark decorations—check. Bizarre and unappetizing camouflage—check.

Let's begin with the pajama squid, which is named for its resting skin pattern of black and white stripes, though like most cephalopods it can change its colors at a moment's notice. (Given their propensity for and skill at disguise, it's surprising that many cephalopods are named for their color or pattern—attributes that seem temporary at best. But often a species has a "normal" or "resting" look that's fairly reliable.) Like other bobtail squid, the pajama squid is incredibly cute. Round body, sassy little fins, big round eyes, tiny arms—it's basically a chibi-style anime character.

Once you take a closer look, though, you can see that the pajama squid is also incredibly gross. Mollusks are, after all, creatures of mucus, and cephalopods take this heritage seriously. Pajama squid secrete a specialized mucus that they use, not to glue themselves to grass as pygmy squid do, but to glue grains of sand into a cozy little snot-house that hides them from predators.

Fish are no slouches when it comes to slime either. The layer of mucus on the skin of many fish can reduce drag and deter parasites, and the reigning monarch of marine slime is also a fish: the hagfish, whose copious secretions clog up the gills of its would-be predators.

Slime aside, let's continue our tour of sepioid oddities. Consider the flamboyant cuttlefish. Painted in vivid reds and yellows, this animal adds to its visual appeal with displays that look like clouds rippling over its body. Why so showy? Like brightly striped caterpillars and spiders, the flamboyant cuttlefish is warning potential predators of its toxic nature. The toxin it houses could easily kill a fish, and may be created by symbiotic bacteria living within the cuttlefish's body.

Evolution has given rise to a number of great bacteria and cephalopod teams, in fact. There's another bobtail squid that farms luminescent bacteria inside an organ uniquely evolved for the task, creating a glow on its belly for countershading camouflage. Like the white, stripe-free underside of a nautilus shell, this bobtail squid's shiny belly blends in with light coming down from the surface, which thereby renders the squid invisible to any creature looking up from below. This intimate squid–bacteria symbiosis has been the focus of extensive research, helping scientists to understand in detail the communications a host can have with its inhabitants. Similar communications take place between our own bodies and the uncounted bacterial species we host. (Our bacteria don't help us light up, because we don't have a particular need for countershading to avoid predators; instead they help us digest our food and protect us from disease.)

Hosting bacteria isn't the only way that cephalopods can make light, though. Many species of deep-sea or midwater squid produce their own luminescence. One of the most spectacular examples is the strobe squid. The flashing photophores on its arms might attract or stun prey, though it's hard to be sure because of the difficulty of studying deep-sea animals in a normal habitat. How can you see what they're doing unless you shine artificial light on them, which will almost certainly alter their behavior? (Still, the luminescent lure of an anglerfish is so obvious in shape and purpose that the animal was named after its ability to "fish" for its prey.)

Curiously, though there are light-up cuttlefish and glow-in-the-dark squid, no octopuses are known to luminesce. This absence of glow may be related to their absence of bone — octopuses are so soft and so squishy that they can squeeze themselves into almost any hole, nook, or cranny. They have perfected the art of hiding from both predators and prey, so they have no use for bright lights. They are ambush predators, their goal not to be seen by their prey until it is already captured and preferably half-eaten.

Evolution has given octopuses a multitude of techniques for hiding. Squirreling down into tiny holes is just one of them. Another is to carry their hiding place around with them, as in the case of the remarkable coconut octopus, which lives in the waters around Southeast Asia. Humans in the area frequently discard empty coconut shells in the water, and octopuses have apparently figured out that these make excellent mobile dens. They'll carry a couple of coconut halves around with them, ready to climb inside and pull the halves together to give the appearance of a whole coconut whenever they need to hide.[10]

Coconut octopuses can make themselves look like coconuts even without the visual aid, by gathering most of their arms around their body to make a rough sphere. They can then balance on just two of their arms and "walk" along the ocean floor, a very un-octopus look that could be mistaken for a coconut rolling along the ground.[11] This is reminiscent of the way a leafy sea dragon imitates algae in both its physical form and its movement.

The mimic octopus — another wonder of the tropical Indo-Pacific — has a far wider repertoire. Members of this species have been seen to imitate flatfish, eels, even a shrimp.

And along with its mimicry, this octopus has another technique for tricking predators: its skin displays a brown-and-white-striped pattern similar to that of the pajama squid. Although such dramatic stripes might seem dangerously noticeable, the fact is that they break up the overall octopus shape. Anyone watching

FIGURE 8.3 A close relative of the mimic octopus
shows off its gaudy stripes. This species is called,
most marvelously, *Wunderpus photogenicus.*
Richard Ross

thinks, "Oh, a collection of lines," instead of "Oh, an octopus," and
thus the creature buys itself time to escape.

So many cephalopod wonders seem to have evolved to keep
these animals from being eaten that it's possible to forget they
themselves are almost universally predatory.

Dangerous Suckers

Cephalopods are, of course, wild animals, and as such should be
treated with respect and caution. They are armed — eight-armed
or ten-armed — and also weaponized with a very sharp beak.
What's more, they have complex brains and have been described
by many aquarists and scientists as curious. So even if they have no
interest in eating you, they may have some interest in *experiencing*

FIGURES 8.4A & 8.4B The vampire squid: from the side (*above*),
and from the mouth (*opposite*). The mouth view displays the soft tendrils
that can easily be mistaken for terrifying spikes. If this book is your own,
or if it belongs to a friend who would enjoy the surprise, you
could color the animal's body red and its eyes bright blue.
Carl Chun, Die Chephalopoden, *1910*

you, as a sort of education. Anyone wishing to understand tales
of squid "attacks" needs to take all of this into account. In reality, no species of squid, vampire or Humboldt or any other, has
ever caused a documented human death. The only cephalopod
that has racked up a body count is the pretty little blue-ringed
octopus.

However, unquestionably the cephalopod with the most fright-
ening name is *Vampyroteuthis infernalis*, which means "vampire
squid from hell." The animal's appearance is also quite spooky. Its
skin is a constant deep red — like many other deep-sea cephalopods, the vampire squid has mostly abandoned its color-changing
abilities as useless in this dark environment. Red is just as good
as black if you want to hide in the deep sea, since red light is absorbed most readily by water and is virtually absent below a few
meters. And then, vampire squid have blue eyes. You might think

these "baby blues" would offset the hellish red, but consider this: the eyes are completely blue — there's a pupil, but you can't see it.

Now add to this the fact that one of the animal's habits is to turn itself partially inside out, wrapping its arms and the webbing between them around its body. The underside of the arms bear rows of sharp-looking tendrils. (They're actually cirri, the same soft skin flaps that give cirrate octopuses their name.) So: it's a red squid with vacant blue eyes that encases itself in apparent spines. We can have some sympathy for the scientists who named it.

But vampire squid are much smaller than you'd fear from photos that never seem to include a scale object. I certainly imagined them to be at least cat-size, until I saw one in person. More like newborn-kitten-size! Even so, until very recently we didn't know what they ate, so you could have held out hope that they would turn out to be actual vampires. Vampire bats are small but still creepy, right? Maybe the vampire squid's two slender filaments could somehow help these cephalopods suck blood.

Nope. In 2012, scientists figured out that the filaments are used to gently filter crud out of the water. Vampire squid, we learned, eat feces and mucus and bits of dead plants and animals and pretty much anything else that falls down to them from the surface waters — a collection of biological trash known euphemistically as "marine snow."[12]

This placid poop-eating lifestyle, though, is unusual among squid, most of whom are keen predators. One species in particular, my science pal the Humboldt squid, has managed to acquire a fearsome name for itself. Though members of this species don't grow as large as giant or colossal squid, the largest adults can still be 6 or 7 feet (about 2 m) long, with the tentacles stretching considerably longer. (Many adults never do get that big, and at certain times in certain places Humboldt squid never grow beyond 2 or 3 feet/0.5 to 1 m, but that's not as exciting.) Their beaks are sharp; their arms are strong; their suckers are equipped with rings of tiny teeth. These sucker rings are more like the hooks of Velcro than like teeth in a predator's jaw; their purpose isn't to damage prey but simply to hook around the edges of fish scales or shrimp skeletons and thereby enhance the arms' grabbing and holding power. This doesn't stop sensationalist news coverage from referring to "squid with ten thousand teeth" or "razor-sharp sucker rings."

Humboldt squid have certainly made divers very nervous, and have even injured them, though probably without any intent to consume them. Unlike great white sharks, which really do confuse surfers with their normal seal prey, Humboldt squid habitually eat small fish and shrimp just a few inches long.[13] While tall tales may circulate about Humboldts eating fishers who have fallen overboard at night, none have ever been substantiated by scientists or journalists.

There is, however, one cephalopod that *has* caused human deaths. The venom of the stunningly gorgeous blue-ringed octopus contains a paralyzing neurotoxin potent enough to kill. This octopus lives, like the flamboyant cuttlefish and so many other beautiful and dangerous creatures, in the tropical Pacific, notably on the Great Barrier Reef.

Also like the flamboyant cuttlefish, the blue-ringed octopus uses bright colors to warn us away, but humans seem not as finely tuned to such warnings as the rest of the animal kingdom. In classic octopus fashion, the blue-ring's shyness usually overpowers its curiosity. Generally the octopus bites only when it's left with no other option for escaping from the big, scary human, which it can only assume is a predator come to devour it. Maybe a person doesn't see the octopus until it's underfoot (blue-ringed octopuses can live in quite shallow water on rocky shores or mudflats). Maybe the person thinks the octopus is pretty and tries to pick it up or (God forbid) bring it home.

On that note, you may occasionally see these creatures for sale in tropical fish stores. Not only would you be courting disaster by bringing such a deadly animal into your own home, but collecting them from the wild may be environmentally damaging, as nobody knows how many there are and whether their populations can sustain collection.

Actually, that last sentence is true for almost every cephalopod alive.

Count 'Em Before You Catch 'Em

Roy Caldwell, a cephalopod biologist at the University of California, Berkeley, has written a succinct plea to hobby aquarists not to buy blue-ringed octopuses, entitled "Blue Ringed Octopus Will Kill You Dead."[14] He points out that the animals are delicate and often die shortly after purchase, if not during shipment. So any living blue-ringed octopus on display bears the hidden burden of several dead, which amplifies the impact on wild populations of even a small number of attempts to keep them.

Equally spectacular but much rarer species like the mimic octopus also tempt aquarists to turn them into pets. But Caldwell reminds us that their habitat in the Southwest Pacific currently suffers from pollution and destructive mining, so perhaps these octopuses have enough problems without being scooped up and shipped around the world to hobbyists.[15]

They might not even make very good pets. As experienced aquarist Christopher Shaw points out, animals that seem flashy when photographed in the wild can turn dull in captivity. "If you want a cool fun octo that will blow your socks off," he writes, "get a bimac."[16] That's an *Octopus bimaculoides*, the California two-spot that makes such a hardy, active, personable pet.

Fortunately, professional aquarists have made tremendous progress in breeding beautiful cephalopod species, so they can be displayed without requiring collection from the wild. While developing the world's largest-ever exclusively cephalopod exhibit, *Tentacles*, researchers at the Monterey Bay Aquarium coaxed flamboyant cuttlefish and pajama squid, among others, to lay fertile eggs that hatched and grew.

As it becomes easier to breed cephalopods, people will inevitably consider farming them for food. But researchers point out that large-scale octopus farming would have significant practical and ethical downsides. Farmers would have to feed each octopus with three times its weight in wild-caught fish, for one thing.[17] And the octopuses would be kept in cramped conditions, just like factory-farmed land animals.

Catching wild cephalopods for food avoids these particular pitfalls. Humans have been catching and eating cephalopods for thousands, maybe tens of thousands of years. But as industrial fishing techniques have replaced artisanal ones, our ability to pull things out of the ocean often outstrips the creatures' ability to replace themselves. Thus, the tragedy of crashing fisheries. We've seen it happen with whales and with cod, with abalone and with lobsters. We haven't seen it happen with any cephalopods — but is that because we're not looking closely enough? Fishery crashes are often visible only in retrospect.

Millions of tons of cephalopods are caught and eaten every year; rarely are any sort of catch limits in place. The California market squid, *Doryteuthis opalescens*, is a notable exception. Though government and academic scientists have been studying the creature for decades, we still don't know how many are out there or how abundantly they reproduce and replace themselves, so the catch limit was created by averaging the years of highest historical catch.

As might be expected from such a limit, it was never reached and therefore fishing was never curtailed. But in 2010, the fishing fleet actually reached the limit before the end of the year, and the season was closed for the first time ever — and many fishers were upset. Since there were more squid available this year than ever before, they argued, shouldn't we increase the limit and keep catching them?[18]

This perspective is not without logic. Squid populations are incredibly variable from year to year, and their spawn-and-die life cycle produces a lot of dead bodies. Why not collect them ourselves rather than go to scavengers or rot? The problem is, you can't catch only squid that have already finished reproducing, nor would you want to — they digest their own bodies as they spawn, and by the time a mother squid has laid her last eggs she's not fit to be a dinner for people. But she's still edible to crabs, worms, sea stars, and many others. In fact, sinking squid carcasses provide an enormous feast for all the deep-sea animals that depend on nutrients raining down from above.[19] Humans fishing for squid inevitably reduce the squids' overall spawning effort, as well as reducing the biomass of dead squid fueling the ecosystem.

So the debate continues. How many squid should we allow ourselves to catch, and how should we change the number as new information comes in and as the population itself changes?

Then there are the cephalopods with charisma. Market squid tug on very few heartstrings. Giant Pacific octopuses, on the other hand, have a fan club. Most of us interact with this species, not in a school dissection lab or on a dinner plate, but in an aquarium setting where they are alive and cared for. Given that, many people are shocked to learn that it's perfectly legal to catch them in the wild and eat them.

In 2013, a furor erupted in Seattle over one teenage diver's hunt for a giant Pacific octopus. He captured the animal by hand in an underwater wrestling match, then dragged it home to eat, but not before a bevy of bystanders had scolded, threatened, and photographed him. Angry letters were written, protests were staged. It turned out that the area where he'd gone hunting was a favorite of scuba divers seeking to view giant octopuses in the wild. It also turned out that he wasn't the only person in Seattle who wanted to eat octopus. A compromise was reached — protect octopuses in that area, because people like to see them there, and let fishing for them continue in other places.[20]

It's not unusual to overhear people guess or assume that the giant Pacific octopus is endangered, simply because it's large and beautiful (a tragic reflection on what it means to be a large, beautiful species in the Anthropocene). But with its prolific procreation, speedy growth, and healthy environments, this species is doing quite well.

Only two cephalopod species are listed as endangered on the IUCN Red List of Threatened Species; both are obscure octopuses that have never been in an aquarium and that nobody ever eats. Their problem is of a different sort. Their problem is *bycatch*.

These octopuses live on seamounts, which are underwater hills whose tops are still deep below the surface — but not deep enough for their denizens to escape the reach of trawl nets. And the octopuses share their seamount homes with several species that people do like to eat, like scampi shrimp and orange roughy fish. The same nets that catch these edible species also happen to collect octopuses, and when the nets are brought to the surface, the undesirables are tossed back into the sea. They almost never survive.[21]

Isolated in the cold depths, these seamount octopuses have evolved away from the "live fast, die young" strategy. Their reproduction is slow and deliberate, producing few eggs, so it's hard for them to recover from any setback.

The IUCN list was only completed in 2014, and it's not clear yet what changes may come of it. Meanwhile, in 2018, the archaic and beautiful pearly nautilus was finally given protection as an endangered species—thanks to years of hard work by scientists, aquarists, conservationists, and an eleven-year-old kid.

Meet the Nautilus (While It's Still Around)

Nautilus meat isn't in high demand; it's fished in small amounts by hungry people who happen to live close to nautiluses, and it's probably not putting too much pressure on nautilus populations. The shell trade, by contrast, is a real doozy. People all over the world love nautilus shells, with good reason. Both the tiger-striped exterior and the mother-of-pearl interior are gorgeous, and the spiral shape gives us a deep aesthetic and mathematical pleasure.

So people buy nautilus shells, original or polished, intact or sliced, or sometimes chopped up to make buttons and earrings. The format doesn't matter—each shell represents the death of an animal. As much as we might wish it to be so, nautilus shells from natural deaths don't wash up on beaches in the quantity and quality that would make it practical to collect them for sale. So hundreds of thousands of nautiluses are killed each year for their shells.

Meanwhile, myriad other human activities threaten their habitat. How many nautiluses suffer from pollution, fishing with explosives, coastline development? It's almost impossible to quantify all the direct and indirect effects. How many individuals are killed outright? How many are simply weakened, succumbing to hunger or disease earlier than they otherwise would have? How many are sterilized, or make fewer babies, as a result of this stress?

The risks seem as clear to Peter Ward, who has worked at the forefront of nautilus science for decades, as they did to eleven-year-old Josiah Utsch, who learned about Ward's work in 2012 and decided to start a campaign called "Save the Nautilus." The organization has raised thousands of dollars to support Ward's research. As concern about nautilus conservation grew, Americans have petitioned their government to propose listing all nautilus species as endangered at the regular international meeting of CITES. The Convention on International Trade in Endangered Species of Wild Fauna and Flora (whew, let's go back to calling it CITES) holds a Conference of the Parties every three years. Despite the petitions, the US government declined to propose a nautilus listing at the 2010 meeting, and demurred again in 2013, citing the lack of information about these animals. But, in what may have surprised cynics as an actual attempt to address the situation, the US government convened a scientific workshop in 2014 focused exclusively on nautiluses.

The workshop report showed drastic declines in nautilus catches over just a few decades in the Philippines, one of the main sources of nautilus shells shipped around the world. Local fishers were well aware of the crash — they remember that a single trap used to catch multiple nautiluses. Now it takes many traps — sometimes more than a hundred — to get a single nautilus.[22]

After publishing the report, the US government solicited recommendations from the public about how to act on this information. Citizens again requested that the government propose protection for all nautilus species. At the same time, mere months before the 2016 CITES meeting, the Center for Biological Diversity petitioned the US National Marine Fisheries Service to list nautiluses under the Endangered Species Act, an action closer to home that could rally support for a CITES decision. In August 2016, the National Marine Fisheries Service agreed that protection *might* be warranted, and they'd take a year to review it. They began with a sixty-day public-comment period.

I learned about this listing through a curious connection: one of the leading scientists working on the effort, Heike Neumeister, is an alumna of the same lab where I did my PhD. We didn't overlap, as she finished her postdoc and moved away before I began my doctorate, but her scientific ghost was present in numerous boxes of laboratory materials and protocols. We got in touch only recently, as I asked her to keep me updated on the nautilus project.

Neumeister's was one of ten submitted public comments that I saw on the document in September. Hers was the longest, a five-paragraph essay outlining the threats nautiluses face (overfishing, pollution, global warming, acidification, even ecotourism) and the value of keeping them around. An excerpt:

First, saving a living fossil from extinction that has survived mostly unchanged for nearly 500 million years. Second, given the obvious importance of the marine ecosystem, maintaining biodiversity in deep reef communities of the tropical Pacific is essential for their existence. Third, Nautilus represents a window to the past from which we can gain valuable scientific insights into biology and ecology of life in ancient oceans; e.g., understanding memory and other processes in an animal that dates back so far in time is a unique opportunity to understand the evolution of the brain including its vulnerability. As such its protection is not only important for us but indeed is a moral imperative for mankind.

She asked me to submit a comment too. I read the "tips for submitting effective comments" on the regulations.gov website. I chewed my fingernails. I reread all the submitted comments. Should I comment at all or stick to uninvolved reporting? If I were to comment, would I wear my cephalopod scientist hat or my science journalist hat? Finally, I composed this:

Nautiluses represent the last remnants of the lineage of shelled cephalopods that once filled the seas; they are intellectually invaluable both for their unique living biology and as a key to the distant past. Furthermore, far from being an evolutionary dead-end, they may well be on the brink of an exciting new species radiation. By listing the nautilus as endangered and preserving it as best we can from the well-documented threat of overfishing, we give ourselves and our children two gifts: the opportunity to enjoy the beauty of living nautiluses in the wild and in aquariums, and the thrill of seeing what evolution does next with these remarkable creatures.

Then, I waited, sitting out the yearlong review process for the Endangered Species Act. Just a few weeks into this wait, however, the United States proposed a nautilus listing at the 2016 CITES meeting. The international vote was overwhelmingly in favor. Nautiluses received their first legal protection.

In 2018, the US fisheries service at last followed suit, giving nautiluses a "threatened" status under the ESA. While it's an encouraging step, they made no corresponding move to place limits on US imports of nautilus shells, pointing out that the CITES listing is already in place to limit exports.

CITES does not forbid the collection or export of nautiluses, but it does assign burden of proof. Any country, like the Philippines, that wishes to export nautiluses must now produce a report showing that the collection isn't harming wild populations — in other words, that the nautilus fishery is sustainable. Scientists like Gregory Barord, who has worked extensively on nautiluses and joined Peter Ward in his rediscovery of *Allonautilus*, are eager to gather data and help create these reports. The ideal outcome would be that we learn enough about nautiluses to implement a sustainable fishery that works for everyone—local fishers, shell lovers, and of course the animals themselves.

"If nautiluses are gone," Barord says, "no one is happy."[23]

Epilogue

Where Are They Going?

The flexibility that has brought cephalopods through numerous mass extinctions — often with drastic changes in form — may yet allow them to rule the seas again. Although nautiluses' long life span and slow rate of reproduction make them particularly vulnerable to overharvesting, most coleoid cephalopods with their "live fast, die young" strategy are particularly resilient. Squid may even be among the animals best able to adapt to a changing planet. Far from being picky eaters, they readily adopt new prey items, and certain species are highly tolerant of warming water with a reduced oxygen concentration.

Alongside the evolutionary innovation of cephalopods, the scientific innovation of humans is ever improving upon the challenges of both marine biology and paleontology. Sophisticated diving robots are learning to explore and sample the most delicate deep-sea creatures and their most unusual behaviors — such as the garbage-slurping vampire squid. Sensitive scanning machines are visualizing ever more tiny and complex structures inside fossils — like the fragments of a baculite's last meal stuck to its radula.

With hundreds of millions of years of cephalopod history behind us, let's peer as best we can into their future. What will the seas of tomorrow look like, and will cephalopods pull through the next mass extinction?

Though the human timescale is evanescent by comparison with the fossil record, we've learned an enormous amount over the past hundred years of cephalopod science — from the prescient visions of Adolf Naef to the latest 3D scans of Isabelle Kruta. What discoveries await in the rock quarry, the laboratory, and the open sea?

Global Proliferation

The world may be seeing a "cephalopod boom," according to evidence amassed in 2016 by Zoë Doubleday of the University of Adelaide and a roster of colleagues, including the irrepressible Sasha Arkhipkin. Their study was prompted by a perception among scientists as well as fishers and other ocean-interested folks that, as Doubleday puts it, "cephalopod populations are proliferating in response to a changing environment."[1]

This makes perfect sense to me, a squid biologist minted in California. Our state's biggest fishery targets market squid, and recent catches have been so high they reached the regulatory limit for the first time in history. Californians have also had front-row seats to the range expansion of the Humboldt squid, once found only as far north as Mexico but reaching the coast of Alaska in 2005. Here on the eastern rim of the Pacific Ocean, we are rather bullish on squid.

But what about *all* the cephalopods in *all* the world?

To tackle that question, Doubleday and her colleagues cast a broad net for data. They cast backward in time to the 1950s and across habitats from the shallows to the depths. They cast wide enough to encompass squid, octopuses, and cuttlefish. They reeled in data both from fisheries and from scientific surveys. Fishery data alone can be misleading when it comes to trends in abundance, because it's hard to tell if fisheries are catching more just because people are trying harder. Maybe the technology became more efficient, or new fishing grounds were discovered. But the survey data corroborated the fishery data, and all the numbers say one thing: cephalopods are trending up.

There's no single explanation for this, but several factors could contribute. One is the plasticity that Yacobucci found in ammonoids. If we think of the evolutionary process as a metaphorical sculptor, then different organisms could be seen as different materials. Some creatures are, like marble, slow to change shape. Cephalopods in this analogy would be more like wet mud, thanks in part to their speedy growth. Already quick to mature, coleoid cephalopods could be growing even more quickly thanks to global warming.

Human activity might give cephalopods another edge as well. The head-footers have been hunted, devoured, and outcompeted by vertebrates ever since fish first evolved; but now big vertebrate predators are the animals that suffer most from overfishing. Bluefin tuna and hammerhead sharks, to name just two, fetch lucrative market prices and are consequently being wiped out of the seas. Cephalopods could conceivably be thanking us for this unintentional release from predation and competition. "So long and thanks for *removing* all the fish!"

The cephalopod boom is great news for all the many predators and scavengers—including humans—that eat cephalopods. It's not such great news for all the animals that are small enough to be eaten by cephalopods instead. And unfortunately, humans are such generalists that we *also* eat a number of these small species, like sardines and shrimp. "Hey squid," some fishers are already thinking, "stay away from our livelihood!"

Doubleday also notes in an award-worthy understatement that "cephalopod population dynamics are notoriously difficult to predict." Cephalopods, being so plastic and adaptable, continuously throw us for a loop. This, too, I have seen in California: two mega-boom years of market squid were followed by unremarkable catches. Half a decade of freaking out over the Humboldt squid invasion was followed by a total absence of these creatures from Californian waters and a distinct reduction in the physical size of these squid in Mexican waters.

Although we've been talking about cephalopods, the population boom belongs exclusively to the coleoids. Nautiloids, alas, are getting the short end of the stick. Though no nautilus species has yet been driven extinct by humans, certain local populations have, and the local populations are too isolated to be replenished, so these extinctions are probably forever. And although the CITES listing will hopefully promote protection from overfishing, nautiluses face nonhuman threats as well. Some of their most common predators, in fact, are the shell-drilling, poison-spitting octopuses. Thriving coleoids would not have the slightest qualm about continuing to devour nautiluses (endangered or not), no more than they would be considerate enough to avoid eating sardines out of our fishing nets.

So the prospect of an ocean overrun with cephalopods, as in the good old Ordovician days, is not unproblematic. The concerns it provokes are reminiscent of worries about a "sea of jelly" that have been stoked by anecdotal evidence of rising jellyfish blooms around the world. Jellies share some qualities with coleoids: they are short-lived adaptable predators that make abundant babies and can therefore quickly take advantage of any environmental change that benefits them. No scientific study like Doubleday's has yet found evidence for a global increase in jellies — due largely to an absence of available data.[2] It's quite possible that jellies will join coleoid cephalopods as the latest in a list of "weedy species" that flourish in the wake of human disturbance. Our species is in the midst of changing Earth in a way that only meteors and volcanoes have managed in the past, and the probable survival of squid and jellies along with rats and mosquitoes is small consolation for the mass exodus of less resilient life forms.[3]

Still, although the death toll humans have racked up over our species' short lifetime makes me wince, the dedication with which many of us have been tackling the problem makes me proud. After all, an international convention just bit into the challenge of keeping nautiluses around. It may be too late to reverse the change, but we can certainly be part of the recovery.

Lab Rats and Libraries

Though Doubleday and her colleagues were the first to quantify the cephalopod boom, numerous other researchers anticipated it. Here's a prescient quote from a 2014 paper by José Xavier of the University of Coimbra, Portugal:

> Cephalopods evolved from an ancestral mollusk in the Cambrian. They have survived major extinction events at the end of the Palaeozoic and at the end of the Mesozoic, and have thrived in spite of competition from fish. Although some cephalopod groups such as ammonites and belemnites became extinct in geological time, the coleoids have survived and radiated. Their life history traits have adapted them for ecological opportunism and provide them with the potential to quickly evolve in response to new selection pressures. There is, therefore, reason to believe that these characteristics will enable cephalopods to evolve under global climate change, enabling them to avoid becoming extinct, and ultimately giving rise to new forms adapted to a new "greenhouse world."[4]

Xavier and his colleagues had convened at a workshop entitled "Future Challenges in Cephalopod Research," and the resultant paper, quoted above, provides a delicious smorgasbord of open questions just waiting for scientists to tackle. Two advances stand out as particularly necessary for furthering our understanding of modern cephalopods: "lab rats" and a library of species.

To understand why, let's start with a puzzle that's almost as old as I am: the paradox of large squid. When the biologist Daniel Pauly sat down in 1988 to calculate exactly how much energy a squid would need to consume in order to get really big, he became convinced that substantial growth would take a long time. And yet calculations of squid age based on layers in statoliths — those little stones in their ears — tell a tale of speedy growth. A Humboldt squid, for example, might grow as long as I am tall in less than a year. (It took me twenty to finally crack 5 feet/1½ m.) Was there an error in Pauly's calculations? Maybe, but no one's been able to find it.

If we could raise these large squid in a laboratory, we could measure how much food they eat and how fast they grow. One way or another, the paradox would be solved. But I banged my head against this challenge for the duration of my PhD, with no success. Squid are just so finicky. Though I kept a pet octopus at the tender age of ten, I have never successfully raised a squid. There are many aquarists far more skilled and more successful than I, like those at the Monterey Bay Aquarium who have raised pygmy squid and reef squid — but even they have not been able to raise large pelagic squid like Humboldts.

In general, squid are harder to raise than octopuses or cuttlefish because they won't settle quietly near the bottom. They want to swim, and then they inevitably bang into the walls of whatever you're keeping them in and damage themselves. Pelagic squid are especially challenging because they're used to a wide-open blue space, which is simply unreplicable in a laboratory. And large pelagic squid like Humboldts are the most difficult of all, because they have the smallest babies. Small babies are fragile, hard to feed, and easy to lose.

If it's difficult to simply keep these animals alive in the laboratory, it might seem like a fool's errand to turn any cephalopod into what biologists call a "model organism" — a species so reliably easy to rear that its biology can be studied in near-infinite detail. Species like fruit flies and lab rats. Still, a cephalopod model organism is one of the scientific developments Xavier and his colleagues call for in their paper, and lo! Eric Edsinger-Gonzales, whom we already know for his role in sequencing the octopus genome, has been working at transforming the adorable pygmy squid into just such a model organism. This species' small size and stationary habit circumvent some of the main challenges of cephalopod rearing.

Pygmy squid, in fact, are about the same size as another aquatic model organism: zebrafish. Zebrafish fill research tanks around the world; they were sent to space in 1976, cloned in 1981, and contribute daily to the latest studies on cancer, birth defects, and more. Pygmy squid and zebrafish have another extremely useful feature in common: transparent eggs and embryos. As you might imagine, transparent organisms are great for doing science. Every step of the animal's development is plain to see, and specific features like the growth of a particular neuron can be easily tracked with fluorescent markers (not the kind you'd highlight a textbook with, but the molecular kind). It's even possible to shine light into transparent organisms to control their neurons, as long as you've spliced in a light-sensitive gene. "You can stop and start a heartbeat with light," says Edsinger-Gonzales. "You can make worms crawl."[5]

With a cephalopod model organism, techniques like this could be applied routinely to an animal that is only very distantly related to us, yet has converged with astonishing detail on many familiar vertebrate features. According to Xavier and colleagues, this "may not only provide further insight into cephalopod evolution, but also into the evolution of man."[6] Knowledge about how eyes, brains, and behavior evolved in the mollusk lineage could offer a unique perspective on how they evolved in our own.

It all sounds like the biology of the future, lab coats and microscopes — far removed from the olden days when naturalists walked in the field with butterfly nets and drew pictures of animals in their notebooks. But as it turns out, even the most cutting-edge science depends on "old school" natural history. Genetics, developmental biology, and the evo-devo work that marries the two all need the context of the whole animal. There's just no substitute for really looking at a creature, understanding and measuring its shape and features in order to see where it fits on the evolutionary tree of life.

But it takes a lot of time and effort to describe, draw, and document the details of a beast. Scientists, always pressed for time with more questions than can possibly be answered in the span of one life, tend to opt out of careful measurements and simply take a DNA sample instead. Grind it up, read a few strands, and you can usually match it to a record online that'll tell you what species you've got. But you're left without the physical details.

New technology may be able to solve this problem, letting us have our cake (our anatomically accurate cake) and eat it too (where "eat" is shorthand for "do lots of experiments that we wouldn't otherwise have time for"). Why couldn't 3D scanners, of the type Ritterbush is using on fossil shells and Kruta is using on fossil radulas, be applied to the tasks of modern morphology? Let's keep up with the technological times, suggests Xavier. Scientists could start publishing, instead of a boring old 2D publication (like, alas, the one you are reading at the moment), a 3D scan of, say, an octopus. "Cephalopods constitute a small-enough class of mollusks that an effort to digitally scan one representative from each genus or species would constitute a realistic goal," writes Xavier.[7]

Imagine a digital cephalopod library that you could browse on your computer or virtual reality headset. You could pace the length of a giant squid, cup a blue-ringed octopus in your hand with no fear. And once there are 3D scans, there can be 3D prints. You could select any cephalopod species to print in your own lab or home to peruse at your leisure — rocketing the dusty old science of taxonomy into the twenty-first century.

Thus, the study of modern living cephalopods meets the study of dead fossilized ones, as computed tomography, 3D printing, and scanning become practical tools for daily research. In one lab, we can imagine Xavier scanning and printing octopuses; in another, Ritterbush scans and prints ammonoid shells. Both avenues open our eyes to the wonder and majesty of the natural world, present and past.

The Future of the Past

When I began delving into the study of extinct cephalopods, I found myself referring again and again to a weighty tome called *Ammonoid Paleobiology*, or, to specialists, simply "The Red Book." This academic doorstopper addressed any question you might care to ask about the most prolific group of ancient cephalopods. But with a publication year of 1996 it was a bit dated, and I often found myself scrambling through more recent literature to find refinements or contradictions of information in The Red Book.

One of the most delightful moments of my research was learning that a new edition was in progress, and as soon as possible I acquired a copy for myself. The 2015 edition of *Ammonoid Paleobiology* had been expanded into two doorstoppers, but fortunately for the sake of my small office there was also a digital edition.[8] The book's weight, however, is metaphorical as much as literal; the new preface opens with this charming analogy: "Imagine you belong to any religion and your chief deity asks you: 'Could you imagine editing the new sacred book?' This is the feeling you have as an ammonoid worker, when you are offered to take care of the new edition of 'Ammonoid Paleobiology.'"

One of the saddest moments of my research came one paragraph later, as the editors of the book paid homage to brilliant minds of earlier times who spent their lives unraveling ammonoid mysteries. "Many important cephalopod workers and good colleagues have passed away in the last two decades," they wrote, listing names that were as familiar to me from ammonoid literature as Tolstoy, Dostoyevsky, and Chekhov would be to a student of Russian literature.

My sorrow over the loss of so many eminent paleontologists, however, is tempered by the infectious enthusiasm of the younger generation. As I discovered one warm evening in Denver, the international community of cephalopod researchers is as vibrant and bustling today as it has ever been in history.

The annual meeting of the Geological Society of America is enormous. Topics cover everything from dinosaurs to oil fields to volcanoes to climate change; in 2016, I was one of seven thousand attendees from forty-eight countries. Despite taking in an abundance of interesting talks by smart scientists (including David "I don't know what causes mass extinctions" Bond), my main purpose was to attend a special submeeting called "Friends of Cephalopods."

Ever since 1976, for one evening during the multiday GSA conference, everyone who loves cephalopods has been getting together to talk shop. Despite a note at the very first meeting stating that the group would need a new name because "Friends of—" had "not been enthusiastically received," the name stuck, and so have the devotees: Gerd Westermann was at the first meeting; Neil Landman has been coming since 1981; Peg Yacobucci since 2000. Yacobucci now organizes and runs the meeting, and Landman's benevolent presence is felt throughout. What everyone was most excited about in 2016, though, was all the new people. With forty-one attendees, this was the largest Friends meeting yet, and the room was filled with students.

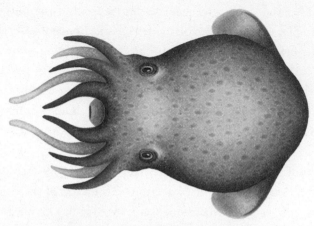

FIGURE 9.1 *Pohlsepia mazonensis* might have been a cute little octopus, as portrayed here, or it might have had little in common with octopuses. As for many fossils, the jury's still out.

Franz Anthony

"It's great for them," someone said, nodding toward Landman and a few other senior scientists, "to see all the students coming into the field."

The wonderful thing about new paleontologists is that they are literally going "into the field," out into the quarries and deserts and riverbeds and cliffs, training new eyes on old rocks. New fossils are bound to be discovered; perhaps one of these students will even find ammonoid soft parts that reveal more secrets about this stupendously successful group.

With advances in technology, new discoveries can be made even with fossils that were dug up decades ago. Sophisticated scanning techniques uncover impressions of soft tissue where none had been visible to the naked eye. This creates a positive feedback loop of more and better data, because as fossil hunters become aware of the available technology, they become more careful in their digging. They bring home extra rock from around the obvious fossil, as what may appear in the field to be merely a gladius could reveal a complete set of arm and sucker impressions in a laboratory scan.

"Now I see the change," says Isabelle Kruta. "I am quite young, but I see that the students and masters students, they do the CT scan now as routine."[9]

In 2016, Kruta and Landman and several others found for the first time a fossilized nerve in an octopus arm.[10] They used an elegant scanning technique called *propagation phase contrast synchrotron X-ray microtomography*, a phrase I plan to use for impressing people if I can ever commit it to memory. In addition to finding the nerve, they were able to count individual suckers and cirri (those little wiggles of skin that look like soft spines). The ability to see this level of detail in fossilized soft tissue opens up the possibility of understanding just how the incredible diversity of modern coleoids evolved. Someday we might be able to test hypotheses about which traits facilitated the shallow-water invasion, for example.

With studies like this coming out every month, I can't wait to see what the future will tell us about the past. Many of the interpretations presented in this book may well prove inaccurate in years to come — as laughable as sluggish dinosaurs with dragging tails and colorless, featherless reptilian skin. But the observations of fossils themselves will long endure, to be reexamined with new tools in new lights. I can't think of a better fate for this book than to amuse my grandchildren, as one of the earliest dinosaur books might elicit chuckles today. After all, those were the books that inspired the next generation of scientists, the next generation of storytellers.

Acknowledgments

Friends of cephalopods—both the forty-one paleontologists who congregated at the 2016 meeting of the Geological Society of America in Denver and the worldwide community of amateur and professional aquarists, rock hounds, and scientists—are some of the kindest people I've met. They're as kind to a ten-year-old who just learned the word "cephalopod" as they are to a graduate student scrambling for data, and as kind again to a writer tackling her first book. I extend eight armloads of gratitude to all the cephalophiles who have inspired and informed me over the years.

Two electronic homes for tentacle science deserve special mention. FASTMOLL is a mailing list for researchers who study "fast mollusks" (which means, of course, the cephalopods), and despite their heavy workloads the scientists on the list responded quickly and generously to all my inquiries. TONMO, *The Octopus News Magazine Online*, is a forum for anyone interested in octopuses (and other cephalopods), amateur or professional. I drank deeply from the resources of this welcoming community.

For sharing their knowledge of the natural world and their enthusiasm for the scientific endeavor, I'm indebted to a great many researchers—I hope I have remembered them all, and I apologize to any whose names I've inadvertently omitted. Kathleen Ritterbush offered prompt assistance with every request, no matter how obscure or mundane, and shared more bizarre and fascinating stories than I could fit in the book. Kenneth De Baets was a tremendously helpful correspondent throughout the research and writing process. Neil Landman and Peg Yacobucci welcomed me warmly into the cephalopod circle, and along with Dirk Fuchs, Christina

Ifrim, and Jakob Vinther, they did their best to straighten out my sketch of the evolutionary tree. Louise Allcock, Sasha Arkhipkin, Matthew Clapham, Jerzy Dzik, Eric Edsinger-Gonzales, Bret Grasse, Kiana Harris, Megan Jensen, Christian Klug, Dieter Korn, Isabelle Kruta, Adrienne Mayor, Heike Neumeister, Scott McKenzie, Neale Monks, Roy Nohra, Andrew Packard, Richard Ross, Isabelle Rouget, Jocelyn Sessa, Martin Smith, and Vojtěch Turek all patiently answered my abundant questions despite the demands of their own work. Many of these generous people also took the time to read and correct portions of the manuscript. Any mistakes that remain despite their efforts are entirely *mea culpa*.

I deeply appreciate the passion that Matthew Lore at The Experiment brought to this project; his enthusiasm bolstered my own. It was his brilliant idea to retitle the book with a nod to the formative influence of Jacques Cousteau. The opportunity to revise, correct, and improve my original text is a precious gift; I hope that I've done it justice.

Editor Liana Willis was a joy to work with, as she guided me along the path toward publication. The book's stunning new cover came from art director Beth Bugler, who also worked with designer Jack Dunnington on the interior. The whole book is so pretty that, at first sight, I nearly squealed.

And, of course, *Monarchs of the Sea* wouldn't exist if *Squid Empire* hadn't been shepherded into the world by a gloriously supportive team. I remain forever grateful for the dedication of agent Lydia Moëd, the advice of editor Stephen Hull, and additional hard work by production editor Ann Brash, copy editor Mary Becker, and designer Eric Brooks. Christine Heinrichs provided helpful comments on an early draft, and Yael Kisel's feedback throughout the writing process was invaluable.

Many thanks to the photographers and artists whose creations illuminate the text; my work would be much poorer without theirs. Cynthia Clark took on the lion's share of the illustration and responded with heroic patience and humor to my every email. The eyes that peer so charmingly from the evolutionary tree of cephalopods are all hers. After the first publication of *Squid Empire*, artist Franz Anthony brought his skill in reconstruction to many of the extinct species that I wrote about, and I'm thrilled to include his illustrations here.

The academic roots of this work owe a great deal to my undergraduate adviser, Armand Kuris, who actively encouraged my octopus obsession, and to my paleontology professor, Bruce Tiffney, who welcomed my term paper about cephalopod evolution — the zeroth draft of this book, you might say. Later support came from my graduate adviser, Bill Gilly, who introduced me to Humboldt squid, taught me sterile laboratory techniques, and unsnarled my fishing lines.

My father's help has been instrumental to the whole journey, from setting up my first aquarium to providing project management as I wrestled with this manuscript. I'm grateful to my children for happily embracing all things tentacled and for challenging me with questions about fossils, extinction, and when Mom will be done writing her book. As for my husband, he brewed pot after pot of tea, took our darling distractions to the park, and cast a critical eye on text and figures. My gratitude to him is as deep as a spirulid's implosion limit — maybe deeper.

Notes

Introduction: Why Squid?

1 Jacques-Yves Cousteau and Philippe Diole, *Octopus and Squid: The Soft Intelligence* (Doubleday, 1973).

2 Here are some of the wonderful cephalopod books of the 2010s that I would have drooled over as a child, and let's be honest: grown-up me still slavered a bit with each publication.

Roland C. Anderson, Jennifer A. Mather, and James B. Wood, *Octopus: The Ocean's Intelligent Invertebrate* (Timber Press, 2010).

Katherine Harmon Courage, *Octopus! The Most Mysterious Creature in the Sea* (Current, 2013).

Sy Montgomery, *The Soul of an Octopus: A Surprising Exploration into the Wonder of Consciousness* (Atria Books, 2015).

Wendy Williams, *Kraken: The Curious, Exciting, and Slightly Disturbing Science of Squid* (Harry N. Abrams, 2011).

3 Jack Prelutsky and Arnold Lobel, *Tyrannosaurus Was a Beast* (Scholastic, 1988).

4 John H. Ostrom, "Osteology of *Deinonychus antirrhopus*, an Unusual Theropod from the Lower Cretaceous of Montana," *Bulletin of the Peabody Museum of Natural History* 30 (1969). 1–165.

5 Robert T. Bakker, "Dinosaur Renaissance," *Scientific American* 232, no. 4 (1975): 58–78.

1. The World of the Head-Footed

1 P. G. Rodhouse, T. R. Arnbom, M. A. Fedack, et al., "Cephalopod Prey of the Southern Elephant Seal, *Mirounga leonina* L.," *Canadian Journal of Zoology* 70 (1992): 1007–1015.

2 I. L. Boyd, T. A. Arnbom, and M. A. Fedak, "Biomass and Energy Consumption of the South Georgia Population of Southern Elephant Seals," in *Elephant Seals: Population Ecology, Behaviour and Physiology*, ed. B.

J. LeBoef and R. M. Laws (University of California Press, 1994), 98–117. South Georgia is an island in the far South Atlantic. If you think of Cape Horn as a tentacle and find a similar tentacle reaching out from Antarctica (the Antarctic Peninsula), the place they're both pointing to is South Georgia. Also, the seals' stomach pumping was preceded by "chemical immobilization," in case you wondered.

3 Food and Agriculture Organization of the United Nations, *FAO Yearbook: Fishery and Aquaculture Statistics* (FAO, 2014).

4 Peter Boyle and Paul Rodhouse, *Cephalopods: Ecology and Fisheries* (Blackwell Science, 2005).

5 Not *all* squid are solid muscle. The jellylike glass squid are mostly water, made even less palatable by a generous helping of ammonia. A variety of marine predators are still willing to chow down on glass squid; humans are not.

6 A. L. Hodgkin and A. F. Huxley, "Action Potentials Recorded from Inside a Nerve Fibre," *Nature* 144, no. 3651 (1939): 710–711.

7 Henk-Jan T. Hoving and B. H. Robison, "Deep-Sea in Situ Observations of Gonatid Squid and Their Prey Reveal High Occurrence of Cannibalism," *Deep Sea Research Part I: Oceanographic Research Papers* 116 (2016): 94–98.

8 The passions that arise over the correct pluralization of "octopus" are truly something to behold. Early in my cephalopod mania I took a strong anti-"octopi" stance, which went along with my generally prescriptivist attitude toward language. I've softened somewhat with age. Here are the best accounts I could find of the etymological history of "octopus": http://www.heracliteanriver.com/?p=240 (accessed October 22, 2016); http://www.grammarphobia.com/blog/2014/02/octopus .html (accessed October 22, 2016).

9 Lawrence Edmonds Griffin, *The Anatomy of Nautilus pompilius*, vol. 5 (Johns Hopkins Press, 1903).

10 Shuichi Shigeno, Sasaki Takenori, and S. von Boletzky, "The Origins of Cephalopod Body Plans: A Geometrical and Developmental Basis for the Evolution of Vertebrate-Like Organ Systems," *Cephalopods: Present and Past* 1 (2010): 23–34.

11 Roger T. Hanlon and John B. Messenger, *Cephalopod Behaviour* (Cambridge University Press, 1998). Despite technically being a textbook, this svelte volume is entirely readable and utterly fascinating. An updated

edition is at the press with a December 2017 publication date, so by the time you read this, it is almost certainly available in bookstores and libraries and I am almost certainly devouring it.

12 Helen Nilsson Sköld, Sara Aspengren, and Margareta Wallin, "Rapid Color Change in Fish and Amphibians: Function, Regulation, and Emerging Applications," *Pigment Cell & Melanoma Research* 26, no. 1 (2013): 29–38.

13 Hannah Rosen, William Gilly, Lauren Bell, et al., "Chromogenic Behaviors of the Humboldt Squid (*Dosidicus gigas*) Studied in Situ with an Animal-Borne Video Package," *Journal of Experimental Biology* 218, no. 2 (2015): 265–275.

14 Ammon got his horns not from an actual ram, but from a merger between foreign deities. The Egyptian god of air was named Amun, "hidden," for his element's invisibility. When the Egyptians conquered Kush, they encountered a major deity with a ram's head and adopted the look for Amun. The Greeks later acquired this god, along with a lot of other Egyptian intellectual property, and misspelled his name.

15 Tingting Yu, Richard Kelly, Lin Mu, Andrew Ross, Jim Kennedy, Pierre Broly, Fangyuan Xia, Haichun Zhang, Bo Wang, and David Dilcher. "An ammonite trapped in Burmese amber." *Proceedings of the National Academy of Sciences* 116, no. 23 (2019): 11345-11350.

16 Neale Monks, Email to the author, May 4, 2016.

17 Neale Monks and Philip Palmer, *Ammonites* (Smithsonian Institution Press, 2002).

18 Ibid., 6.

19 Neale Monks, "Ammonite Wars," *Deposits Magazine*, February 2, 2016, https://depositsmag.com/2016/02/25/ammonite-wars/ (accessed January 28, 2017).

20 Neil Shubin, *Your Inner Fish* (Pantheon, 2008). This is an outstanding introduction to evo-devo as well as an illuminating account of the evolutionary machinations underpinning our own anatomy. The best part: after reading this book, I finally understand hiccups.

21 A brief and lovely account of the volatile history behind the theory of plate tectonics (and its first incarnation, continental drift) can be found in the University of California, Berkeley's biography of Alfred Wegener: http://www.ucmp.berkeley.edu/history/wegener.html (accessed January 28, 2017).

22 David J. Bottjer, "Paleogenomics and Plate Tectonics: Revolutions in the Earth and Biological Sciences," Paper presented at the GSA Annual Meeting, Denver, December 25–28, 2016.

23 Eric T. Domyan, Zev Kronenberg, Carlos R. Infante, et al., "Molecular Shifts in Limb Identity Underlie Development of Feathered Feet in Two Domestic Avian Species," *eLife* 5 (2016): e12115.

24 Peg Yacobucci, Phone interview with the author, April 1, 2016.

25 David M. Raup and J. John Sepkoski, Jr., "Mass Extinctions in the Marine Fossil Record," *Science* 215, no. 4539 (1982): 1501–1503. This is *the* paper that identified what are now called the "Big Five" mass extinctions in the history of life. Raup and Sepkoski identified these extinctions on the basis of patterns Sepkoski had discovered in marine fossils — with an abundant data contribution from cephalopods, of course.

2. Rise of the Empire

1 Caroline B. Albertin, Oleg Simakov, Therese Mitros, et al., "The Octopus Genome and the Evolution of Cephalopod Neural and Morphological Novelties," *Nature* 524, no. 7564 (2015): 220–224.

2 Norman Bertram Marshall, *Explorations in the Life of Fishes*, Harvard Books in Biology, no. 7 (Harvard University Press, 1971).

3 Eric Edsinger-Gonzales, Skype interview with the author, October 5, 2016.

4 Allen P. Nutman, Vickie C. Bennett, Clark R. L. Friend, et al., "Rapid Emergence of Life Shown by Discovery of 3,700-Million-Year-Old Microbial Structures," *Nature* 537, no. 7621 (2016): 535–538. It's worth noting that before this study was published in 2016, the oldest known fossils were a mere 3.4 billion years old. These 3.7-billion-year-old fossils were discovered *only* because enough ice had melted off Greenland to expose them.

5 Peter Ward and Joe Kirschvink, *A New History of Life: The Radical New Discoveries about the Origins and Evolution of Life on Earth* (Bloomsbury, 2015). The idea of proto–living molecules or actual cells arriving on Earth from extraterrestrial sources is called *panspermia* and is enjoyably detailed in this book. Since its publication, even more evidence has cropped up, in the form of complex organic molecules found in the *Rosetta* mission's pursuit of a comet. The book also contains an intriguing discussion of the early appearance and evolution of sponges.

6 Mikhail A. Fedonkin and Benjamin M. Waggoner, "The Late Precambrian Fossil *Kimberella* Is a Mollusc-Like Bilaterian Organism," *Nature* 388, no. 6645 (1997): 868–871.

7 Douglas Erwin and James Valentine, *The Cambrian Explosion* (W. H. Freeman, 2013).

8 Douglas H. Erwin, Marc Laflamme, Sarah M. Tweedt, et al., "The Cambrian Conundrum: Early Divergence and Later Ecological Success in the Early History of Animals," *Science* 334, no. 6059 (2011): 1091–1097.

9 Mark A. S. McMenamin, "The Garden of Ediacara," *Palaios* (1986): 178–182.

10 Predation in its broadest sense, however, has pretty much always been around. As soon as there were cells, they most likely tried to eat each other.

11 Hong Hua, Brian R. Pratt, and Lu-Yi Zhang, "Borings in *Cloudina* Shells: Complex Predator–Prey Dynamics in the Terminal Neoproterozoic," *Palaios* 18, no. 4–5 (2003): 454–459.

12 You'd be quite right to point out that true snails like the ones in your garden also bear single shells. And snails have at times been called univalves — derived from the Latin for "single shell." However, the more common scientific term for snails proper is "gastropod," based on the Greek and referring to their "stomach-on-top-of-foot" architecture.

13 Winston F. Ponder and David R. Lindberg (eds.), *Phylogeny and Evolution of the Mollusca* (University of California Press, 2008). The abundance of useful information in this textbook includes Gerhard Haszprunar's quoted comment on the monoplacophoran "sensation." I won my copy of the book in a haiku contest at the 2008 meeting of the Western Society of Naturalists, which is a serious scientific meeting and is also the kind of meeting that has haiku contests. My winning poem: "Rising from a past / of mucus and cilia / our squid overlords."

14 Björn Kröger, "Comments on Ebel's Benthic-Crawler Hypothesis for Ammonoids and Extinct Nautiloids." *Paläontologische Zeitschrift* 75, no. 1 (2001): 123–125.

15 Curiously, the exact position of the siphuncle varies. Ammonoid and coleoid siphuncles run along the outer rim of the chambers, whereas nautiloid siphuncles run right through the chamber middle. Why? No one knows.

16 M. J. Wells and R. K. O'Dor, "Jet Propulsion and the Evolution of the Cephalopods," *Bulletin of Marine Science* 49, no. 1–2 (1991): 419–432.

17 Martin R. Smith, Skype interview with the author, May 11, 2016.

18 Martin R. Smith and Jean-Bernard Caron, "Primitive Soft-Bodied Cephalopods from the Cambrian," *Nature* 465, no. 7297 (2010): 469–472.

19 Dawid Mazurek and Michał Zatoń, "Is *Nectocaris pteryx* a Cephalopod?" *Lethaia* 44, no. 1 (2011): 2–4.

20 B. Runnegar, "Once Again: Is *Nectocaris pteryx* a Stem Group Cephalopod?" *Lethaia*, 44, no. 4 (2011): 373.

21 Björn Kröger, Jakob Vinther, and Dirk Fuchs, "Cephalopod Origin and Evolution: A Congruent Picture Emerging from Fossils, Development and Molecules," *Bioessays* 33, no. 8 (2011): 602–613. This paper has been cited 110 times according to Google Scholar, which (if the citations are distributed evenly over time) means that every month, one or two papers come out citing Kröger et al. That's a pretty good clip!

22 Peter Douglas Ward, *In Search of Nautilus: Three Centuries of Scientific Adventures in the Deep Pacific to Capture a Prehistoric Living Fossil* (Simon & Schuster, 1988).

23 Neale Monks and Philip Palmer, *Ammonites* (Smithsonian Institution Press, 2002), 55.

24 Neale Monks, "A Broad Brush History of the Cephalopoda," *The Cephalopod Page*, http://www.thecephalopodpage.org/evolution.php (accessed January 20, 2017).

25 Trilobites also evolved drifting and swimming forms in the Ordovician; cephalopods were not the only invertebrates roaming free in the water. The eminent British trilobitologist Richard Fortey has even discovered "streamlined trilobites that sped through the Ordovician ocean," and I wonder if cephalopod predators constituted the evolutionary pressure toward faster swimming in these crunchy little morsels. I've had to give trilobites sadly short shrift, but for the interested reader I highly recommend Fortey's engaging book, *Trilobite! Eyewitness to Evolution* (Alfred A. Knopf, 2000).

26 Christian Klug, Kenneth De Baets, Björn Kröger, et al., "Normal Giants? Temporal and Latitudinal Shifts of Palaeozoic Marine Invertebrate Gigantism and Global Change," *Lethaia* 48, no. 2 (2015): 267–288.

27 Dieter Korn, Skype interview with the author, January 29, 2016.

28 Christian Klug, Phone interview with the author, January 15, 2016.

29 Sarah E. Gabbott, "Orthoconic Cephalopods and Associated Fauna from the Late Ordovician Soom Shale Lagerstatte, South Africa," *Palaeontology* 42, no. 1 (1999): 123–148.

30 There's "ceras" again, referring to the horn shape of the shell. After I complained in *Squid Empire* that I couldn't discover the meaning of "Sphoo," classics professor Dianna Rhyan wrote to inform me, "The *spho-* root is a pronoun or possessive adjective root pertaining to 'two' or 'both' . . . How pertinent this is to the fossil I will leave you to decide." It's perfectly pertinent, as paleontologists pieced together the story of *Sphooceras* from two kinds of fossils: the truncated shells of living animals, and the deciduous shells they left behind. (email to the author, October 29, 2019)

31 J. Barrande, "Troncature normale ou périodique de la coquille dans certains céphalopodes paléozoïques," *Bulletin de la Société Géologique de France*, séries 2, 17 (1860): 573–601.

32 J. Dzik, "Phylogeny of the Nautiloidea," *Palaeontologia Polonica* 45 (1984): 1–255.

33 Vojtěch Turek and Štěpán Manda, "'An Endocochleate Experiment' in the Siluran Straight-Shelled Cephalopod *Sphooceras*," *Bulletin of Geosciences* 87, no. 4 (2012): 767–813. Nobody, including Jerzy Dzik, has refuted Turek and Manda's hypothesis in print, and several references have found it supported by the available evidence. However, when I asked Dzik for his opinion, he noted that mollusk shells of all kinds typically have growth lines, rather like the rings of a tree. To his eyes, the pattern on *Sphooceras*'s shell cap resembles these growth lines rather than a design laid down later by the animal's body. "Ironically, the color pattern observed by Turek and Manda can be used as a support to the idea that the large tip of *Sphooceras* was a protoconch [the earliest part of a mollusk's shell, grown while the animal is still in its egg]!" Dzik wrote. "I do not claim that this is true but the new evidence actually obliterates the issue instead of clarifying it" (email to the author, February 9, 2016).

34 Dzik, Email to the author, February 9, 2016.

3. A Swimming Revolution

1 Christian Klug, Bjoern Kroeger, Wolfgang Kiessling, et al., "The Devonian Nekton Revolution," *Lethaia* 43, no. 4 (2010): 465–477.

2 Christian Klug, Phone interview with the author, January 15, 2016.

3 Christian Klug, Linda Frey, Dieter Korn, et al., "The Oldest Gondwa-
nan Cephalopod Mandibles (Hangenberg Black Shale, Late Devonian)
and the Mid Palaeozoic Rise of Jaws," *Palaeontology* 59, no. 5 (2016):
611–629.

4 Klug, Phone interview with the author, January 15, 2016.

5 Jakob Vinther, Skype interview with the author, March 15, 2016.

6 Dieter Korn, Skype interview with the author, January 29, 2016.

7 Neil H. Landman, William A. Cobban, and Neal L. Larson, "Mode of
Life and Habitat of Scaphitid Ammonites," *Geobios* 45, no. 1 (2012):
87–98.

8 Mark Norman, *Cephalopods: A World Guide* (ConchBooks, 2000). This
guidebook is packed with gorgeous photographs and includes most of
the great cephalopod sex stories: "Breeding Nautiluses," "Giant Cuttlefish Spawning Grounds," "Cross-Dressing Cuttlefish" (that's those
sneaker males), "Sperm Wars," "The Night of Love and Death" (about
mass spawning and subsequent death in squid), "Giant Squid Sex,"
"Eggs, Brooding and Spawning," and "Weird Sex" (including that of the
argonauts).

9 Steve Etches, Jane Clarke, and John Callomon, "Ammonite Eggs and
Ammonitellae from the Kimmeridge Clay Formation (Upper Jurassic)
of Dorset, England," *Lethaia* 42, no. 2 (2009): 204–217.

10 Royal H. Mapes and Alexander Nuetzel, "Late Palaeozoic Mollusc Reproduction: Cephalopod Egg-Laying Behavior and Gastropod Larval
Palaeobiology," *Lethaia* 42, no. 3 (2009): 341–356.

11 Aleksandr A. Mironenko and Mikhail A. Rogov, "First Direct Evidence
of Ammonoid Ovoviviparity," *Lethaia* (2015): 245–260.

12 Peg Yacobucci, Phone interview with the author, April 1, 2016.

13 Kenneth De Baets, Christian Klug, Dieter Korn, and Neil H. Landman,
"Early Evolutionary Trends in Ammonoid Embryonic Development,"
Evolution 66, no. 6 (2012): 1788–1806.

14 The history behind the "midnight tank escape" story is detailed in this
wonderfully thorough thread on TONMO.com: https://www.tonmo
.com/threads/midnight-tank-escapes-fact-or-fiction.16560/ (accessed
January 28, 2017).

15 Brian Switek, *My Beloved Brontosaurus: On the Road with Old Bones,
New Science, and Our Favorite Dinosaurs* (Macmillan, 2013).

16 Robert J. Diaz and Rutger Rosenberg, "Spreading Dead Zones and Consequences for Marine Ecosystems," *Science* 321, no. 5891 (2008): 926–929.

17 David Bond, Paul B. Wignall, and Grzegorz Racki, "Extent and Duration of Marine Anoxia During the Frasnian–Famennian (Late Devonian) Mass Extinction in Poland, Germany, Austria and France," *Geological Magazine* 141, no. 2 (2004): 173–193.

18 Robert Lemanis, Dieter Korn, Stefan Zachow, et al., "The Evolution and Development of Cephalopod Chambers and Their Shape," *PloS One* 11, no. 3 (2016): e0151404.

19 Korn, Skype interview with the author, January 29, 2016.

20 David P. G. Bond and Paul B. Wignall, "Large Igneous Provinces and Mass Extinctions: An Update," *Geological Society of America Special Papers* 505 (2014): SPE505-02.

21 Matthew Clapham, Interview with the author, March 24, 2016.

22 For more details about scientific studies on current and projected ocean acidification, visit this beautifully readable site maintained by the Smithsonian Museum's Ocean Portal: http://ocean.si.edu/ocean-acidification (accessed January 28, 2017).

23 Korn, Skype interview with the author, January 29, 2016.

4. The Protean Shell

1 R. Granot, "Palaeozoic Oceanic Crust Preserved Beneath the Eastern Mediterranean," *Nature Geoscience* 9 (2016): 701–705. The Mediterranean was named "center of the earth" by ancients with an incomplete knowledge of geography, but this name does reflect with a certain charming insularity that the sea is one of the oldest on the planet. Its exact age is still under debate — possibly everything that was left of the original Tethys was subducted under the earth's crust, covered over with a newer (but still old!) ocean called Neo-Tethys, which eventually became the Med.

2 Brad A. Seibel, Fabienne Chausson, Francois H. Lallier, et al., "Vampire Blood: Respiratory Physiology of the Vampire Squid (Cephalopoda: Vampyromorpha) in Relation to the Oxygen Minimum Layer," *Experimental Biology Online* 4, no. 1 (1999): 1–10; Brad A. Seibel, N. Sören Häfker, Katja Trübenbach, et al., "Metabolic Suppression During

Protracted Exposure to Hypoxia in the Jumbo Squid, *Dosidicus gigas*, Living in an Oxygen Minimum Zone," *Journal of Experimental Biology* 217, no. 14 (2014): 2555–2568.

3   The idea of ancient nautiloids seizing the first available oxygen comes from Peter Ward and Joe Kirschvink, *A New History of Life: The Radical New Discoveries about the Origins and Evolution of Life on Earth* (Bloomsbury, 2015).

4   Tamaki Sato and Kazushige Tanabe, "Cretaceous Plesiosaurs Ate Ammonites," *Nature* 394, no. 6694 (1998): 629–630; Judy A. Massare and Heather A. Young, "Gastric Contents of an Ichthyosaur from the Sundance Formation (Jurassic) of Central Wyoming," *Paludicola* 5, no. 1 (2005): 20–27.

5   Erle G. Kauffman, "Mosasaur Predation on Upper Cretaceous Nautiloids and Ammonites from the United States Pacific Coast," *Palaios* 19 (2004): 96–100.

6   Kenneth De Baets, Skype interview with the author, July 24, 2015.

7   Geerat J. Vermeij, "The Mesozoic Marine Revolution: Evidence from Snails, Predators and Grazers," *Paleobiology* 3, no. 3 (1977): 245–258.

8   Peg Yacobucci, Phone interview with the author, April 1, 2016.

9   For the academic paper in which Yacobucci discusses this topic in more detail, see Margaret M. Yacobucci, "Plasticity of Developmental Timing as the Underlying Cause of High Speciation Rates in Ammonoids," in *Advancing Research on Living and Fossil Cephalopods*, ed. Federico Olóriz and Francisco J. Rodríguez-Tovar (Springer, 1999), 59–76.

10 Margaret M. Yacobucci, "An Example from the Cenomanian Western Interior Seaway," In *Advancing Research on Living and Fossil Cephalopods: Development and Evolution Form, Construction, and Function Taphonomy, Palaeoecology, Palaeobiogeography, Biostratigraphy, and Basin Analysis*, ed. Federico Olóriz and Francisco J. Rodríguez-Tovar (Springer, 2013), 59.

11 Kathleen Ritterbush, Skype interview with the author, August 28, 2015.

12 Larisa A. Doguzhaeva and Harry Mutvei, "The Additional External Shell Layers Indicative of 'Endocochleate Experiments' in Some Ammonoids," in *Ammonoid Paleobiology: From Anatomy to Ecology*, ed. Christian Klug, Dieter Korn, Kenneth De Baets, et al. (Springer, 2015), 585–609. This is one of the papers that accepts Turek and Manda's hypothesis about *Sphooceras* and uses it to look at potentially similar experiments in later ammonoids.

13 Yacobucci, Phone interview with the author, April 1, 2016.

14 Kathleen A. Ritterbush and David J. Bottjer, "Westermann Morphospace Displays Ammonoid Shell Shape and Hypothetical Paleoecology," *Paleobiology* 38, no. 3 (2012): 424–446.

15 K. A. Ritterbush, R. Hoffmann, A. Lukeneder, and K. De Baets, "Pelagic Palaeoecology: The Importance of Recent Constraints on Ammonoid Palaeobiology and Life History," *Journal of Zoology* 292, no. 4 (2014): 229–241.

16 Neale Monks and Philip Palmer, *Ammonites* (Smithsonian Institution Press, 2002), 93.

17 Neil H. Landman, J. Kirk Cochran, Neal L. Larson, et al., "Methane Seeps as Ammonite Habitats in the US Western Interior Seaway Revealed by Isotopic Analyses of Well-Preserved Shell Material," *Geology* 40, no. 6 (2012): 507–510.

18 Jocelyn Anne Sessa, Ekaterina Larina, Katja Knoll, et al., "Ammonite Habitat Revealed via Isotopic Composition and Comparisons with Co-occurring Benthic and Planktonic Organisms," *Proceedings of the National Academy of Sciences* 112, no. 51 (2015): 15562–15567.

19 Neale Monks and Jeremy R. Young, "Body Position and the Functional Morphology of Cretaceous Heteromorph Ammonites," *Palaeontologia Electronica* 1, no. 1 (1998): 15.

20 Alexander I. Arkhipkin, "Getting Hooked: The Role of a U-Shaped Body Chamber in the Shell of Adult Heteromorph Ammonites," *Journal of Molluscan Studies* (2014): eyu019.

21 Alexander Arkhipkin, Skype interview with the author, March 14, 2016.

22 Neil H. Landman, Isabelle Kruta, John S. S. Denton, and J. Kirk Cochran, "Getting Unhooked: Comment on the Hypothesis That Heteromorph Ammonites Were Attached to Kelp Branches on the Sea Floor, as Proposed by Arkhipkin (2014)," *Journal of Molluscan Studies* 82, no. 2 (2016): 351–355.

23 Alexander I. Arkhipkin, "If Not Getting Hooked, Why Make One? Response to Landman et al.," *Journal of Molluscan Studies* (2016): eyv065.

24 Neil H. Landman, William A. Cobban, and Neal L. Larson, "Mode of Life and Habitat of Scaphitid Ammonites," *Geobios* 45, no. 1 (2012): 87–98, at 93.

25 U. Lehmann, "Ammonite Jaw Apparatus and Soft Parts," *Ammonoidea: The Systematics Association Special* 18 (1981): 275–287.

26 Horacio Parent, Gerd E. G. Westermann, and John A. Chamberlain, "Ammonite Aptychi: Functions and Role in Propulsion," *Geobios* 47, no. 1 (2014): 45–55.

27 Isabelle Kruta, Skype interview with the author, March 31, 2016.

28 Isabelle Kruta, Neil Landman, Isabelle Rouget, et al., "The Role of Ammonites in the Mesozoic Marine Food Web Revealed by Jaw Preservation," *Science* 331, no. 6013 (2011): 70–72.

29 K. N. Nesis, "On the Feeding and Causes of Extinction of Certain Heteromorph Ammonites," *Paleontological Journal* (1986): 5–11.

30 Jakob Vinther, Skype interview with the author, March 15, 2016.

5. Sheathing the Shell

1 A. Packard, "Operational Convergence Between Cephalopods and Fish: An Exercise in Functional Anatomy," *Archivio Zoologico Italiano* 51 (1966): 523–542.

2 P. Doyle and D. I. M. Macdonald, "Belemnite Battlefields," *Lethaia* 26 (1993): 65–80.

3 Mico Tatalovic, "Drawing with Ancient Ink," *Nature News*, August 19, 2009.

4 Quanguo Li, Ke-Qin Gao, Jakob Vinther, et al., "Plumage Color Patterns of an Extinct Dinosaur," *Science* 327, no. 5971 (2010): 1369–1372.

5 Jakob Vinther, Skype interview with the author, March 15, 2016.

6 Dirk Fuchs, Sigurd von Boletzky, and Helmut Tischlinger, "New Evidence of Functional Suckers in Belemnoid Coleoids (Cephalopoda) Weakens Support for the 'Neocoleoidea' Concept," *Journal of Molluscan Studies* 76, no. 4 (2010): 404–406.

7 As a curious side note, some male scaphite ammonoids seem to have hooklike structures preserved inside their shells, which Landman has interpreted as remnants of an oversized hectocotylus that might have used such hooks to aid in sperm transfer.

9 Alexandra C. N. Kingston, Alan M. Kuzirian, Roger T. Hanlon, and Thomas W. Cronin, "Visual Phototransduction Components in Cephalopod Chromatophores Suggest Dermal Photoreception," *Journal of Experimental Biology* 218, no. 10 (2015): 1596–1602.

10 Alexander L. Stubbs and Christopher W. Stubbs, "Spectral Discrimination in Color Blind Animals via Chromatic Aberration and Pupil

Shape," *Proceedings of the National Academy of Sciences* 113, no. 29 (2016): 8206–8211.

11 Nadav Shashar, P. Rutledge, and T. Cronin, "Polarization Vision in Cuttlefish in a Concealed Communication Channel?" *Journal of Experimental Biology* 199, no. 9 (1996): 2077–2084.

12 Royal H. Mapes, Larisa A. Doguzhaeva, Harry Mutvei, et al., "The Oldest Known (Lower Carboniferous-Namurian) Protoconch of a Rostrum-Bearing Coleoid (Cephalopoda) from Arkansas, USA: Phylogenetic and Paleobiologic Implications," in "Proceedings of the 3rd International Symposium, 'Coleoid Cephalopods Through Time,' Dirk Fuchs (editor), Luxembourg October 8–11, 2008," *Travaux scientifiques du Musée national d'histoire naturelle Luxembourg* (2010): 114–125.

13 Two of Naef's iconic monographs, on fossil cephalopods and on cephalopod development, which are still regularly referenced by scientists today: *Die fossilen Tintenfische: Eine paläozoologische Monographie* (Fischer, 1922) and "Die Cephalopoden (Embryologie)," *Fauna Flora Golf Neapel* 35, no. 2 (1928): 1–357.

14 Christian Klug, Günter Schweigert, Dirk Fuchs, et al., "Adaptations to Squid-Style High-Speed Swimming in Jurassic Belemnitids," *Biology Letters* 12, no. 1 (2016). http://rsbl.royalsocietypublishing.org/content /12/1/20150877 (accessed January 28, 2017).

15 Ibid.

16 Ibid.

17 Dirk Fuchs, Skype interview with the author, January 18, 2016.

18 Dominique Jenny, Dirk Fuchs, Alexander I. Arkhipkin, Rolf B. Hauff, Barbara Fritschi, and Christian Klug. "Predatory behaviour and taphonomy of a Jurassic belemnoid coleoid (Diplobelida, Cephalopoda)." *Scientific Reports* 9, no. 1 (2019): 7944.

19 Neale Monks and S. Wells, "A New Record of the Eocene Coleoid *Spirulirostra anomala* (Mollusca: Cephalopoda) and Its Relationships to Modern Spirula," *Tertiary Research* 19 (2000): 47–52

20 With fellow squid scientists in graduate school, I started a program called "Squid4Kids," bringing excess catch from squid fishers into schools for students to learn from: http://gillylab.stanford.edu /outreach.html (accessed January 28, 2017).

21 Bruce Hopkins and S. V. Boletzky, "The Fine Morphology of the Shell Sac in the Squid Genus *Loligo* (Mollusca: Cephalopoda): Features of a Modified Conchiferan Program," *Veliger* 37 (1994): 344–357.

22 Dirk Fuchs and Iba Yasuhiro, "The Gladiuses in Coleoid Cephalopods: Homology, Parallelism, or Convergence?" *Swiss Journal of Palaeontology* 134, no. 2 (2015): 187–197, at 187.

23 Mark Sutton, Catalina Perales Raya, and Isabel Gilbert, "A Phylogeny of Fossil and Living Neocoleoid Cephalopods," *Cladistics* (2015): 1–11.

24 Brad A. Seibel, Erik V. Thuesen, and James J. Childress, "Flight of the Vampire: Ontogenetic Gait-Transition in *Vampyroteuthis infernalis* (Cephalopoda: Vampyromorpha)," *Journal of Experimental Biology* 201, no. 16 (1998): 2413–2424. This lovely paper details the bizarre metamorphosis of vampire squid fins. These animals begin life with one pair of fins, then grow an entirely separate and differently shaped pair of fins. Brad Seibel and his colleagues suggest the change is due to a shift in gait as the animal develops.

25 Dirk Fuchs, Christina Ifrim, and Wolfgang Stinnesbeck, "A New *Palaeoctopus* (Cephalopoda: Coeloidea) from the Late Cretaceous of Vallecillo, North-Eastern Mexico, and Implications for the Evolution of Octopoda," *Palaeontology* 51, no. 5 (2008): 1129–1139.

26 Romain Jattiot, Arnaud Brayard, Emmanuel Fara, and Sylvain Charbonnier, "Gladius-Bearing Coleoids from the Upper Cretaceous Lebanese *Lagerstätten*: Diversity, Morphology, and Phylogenetic Implications," *Journal of Paleontology* 89, no. 1 (2015): 148–167.

27 H. Woodward, "On a New Genus of Fossil 'Calamary' from the Cretaceous Formation of Sahel Alma, near Beirut, Lebanon, Syria," *Geological Magazine*, new series, 10 (1883): 1–5.

28 J. Roger, "Le plus ancien Céphalopode Octopode fossile connu: Palaeoctopus newboldi (Sowerby 1846) Woodward," *Bulletin mensuel de la Société linnéenne de Lyon* 13, no. 9 (1944): 114–118.

29 Roy Nohra, Email to the author, October 30, 2016.

30 The Expo Hakel museum website is http://www.expohakel.com/.

31 Adrienne Mayor, *Fossil Legends of the First Americans* (Princeton University Press, 2007), 226–229.

32 Adrienne Mayor, "Fossils in Native American Lands: Whose Bones, Whose Story? Fossil Appropriations Past and Present," Paper presented at the History of Science Society annual meeting, November 1–2, 2007, Washington, DC.

33 Adrienne Mayor, *The First Fossil Hunters: Dinosaurs, Mammoths, and Myth in Greek and Roman Times* (Princeton University Press, 2011).

34 According to the online store Rudraksha from Nepal, "Whether one has real devotion or not, if he worships a saligrama-stone with devotion before it, he will surely [be] liberated from the cycle of phenomenal existence. The person who offers a daily service for the saligrama stone will be freed from the fear of death, and he will cross over the stream of births and deaths." http://www.rudrakshanepal.com/page-36-About_Saligram (accessed January 30, 2017).

35 Michael G. Bassett, *Formed Stones, Folklore and Fossils*, Geological Series, no. 1 (National Museum of Wales, 1982).

36 Adrienne Mayor, "Dinosaurs with Native American Names." *Wonders & Marvels*, 2015. http://www.wondersandmarvels.com/2015/08/dinosaurs-with-native-american-names.html (accessed January 28, 2017).

6. Fall of the Empire

1 L. W. Alvarez, W. Alvarez, F. Asaro, and H. V. Michel, "Extraterrestrial Cause for the Cretaceous-Tertiary Extinction," *Science* 208 (1980): 1095–1108. This is the explosive paper in which Walter and Luis Alvarez proposed that a meteor impact ended the dinosaurs (and ammonoids). For an outstanding breakdown of the turmoil and vitriol and wound licking that followed, read this account compiled by the science writer Ann Finkbeiner: http://www.lastwordonnothing.com/2013/11/11/what-luis-alvarez-did/ (accessed January 28, 2017).

2 Jost Wiedmann and Jürgen Kullman, "Crises in Ammonoid Evolution," in *Ammonoid Paleobiology*, ed. Neil H. Landman, Kazushige Tanabe, and Richard Arnold Davis (Springer, 1996), 795–813.

3 Matthew Clapham, Interview with the author, March 24, 2016.

4 Neil H. Landman, Stijn Goolaerts, John W. M. Jagt, et al., "Ammonites on the Brink of Extinction: Diversity, Abundance, and Ecology of the Order Ammonoidea at the Cretaceous/Paleogene (K/Pg) Boundary," in *Ammonoid Paleobiology: From Macroevolution to Paleogeography*, ed. Christian Klug, Dieter Korn, Kenneth De Baets, et al. (Springer, 2015), 497–553.

5 David P. G. Bond, "The Causes of Mass Extinctions: How Can We Better Understand How, Why and When Ecosystems Collapse?" Paper presented at the GSA Annual Meeting, Denver, September 25–28, 2016.

6 Scientists estimate that both the Siberian and the Indian flood basalts once covered far greater areas — about 580,000 square miles (1,500,000 km²) in India and up to nearly 3,000,000 square miles (7,000,000 km²) in Siberia.

7 "Did Dinosaur-Killing Asteroid Trigger Largest Lava Flows on Earth?" http://www.sciencenewsline.com/news/2015050109530049.html (accessed November 12, 2016).

8 Peter Schulte, Laia Alegret, Ignacio Arenillas, et al., "The Chicxulub Asteroid Impact and Mass Extinction at the Cretaceous–Paleogene Boundary," *Science* 327, no. 5970 (2010): 1214–1218.

9 Kathleen Ritterbush, Skype interview with the author, August 28, 2015.

10 Jocelyn Sessa, Skype interview with the author, January 21, 2016.

11 Neil H. Landman, Stijn Goolaerts, John W. M. Jagt, et al., "Ammonite Extinction and Nautilid Survival at the End of the Cretaceous," *Geology* 42, no. 8 (2014): 707–710.

12 Ibid., 709.

13 Laia Alegret, Ellen Thomas, and Kyger C. Lohmann, "End-Cretaceous Marine Mass Extinction Not Caused by Productivity Collapse," *Proceedings of the National Academy of Sciences* 109, no. 3 (2012): 728–732.

14 Alexander I. Arkhipkin and Vladimir V. Laptikhovsky, "Impact of Ocean Acidification on Plankton Larvae as a Cause of Mass Extinctions in Ammonites and Belemnites," *Neues Jahrbuch für Geologie und Paläontologie-Abhandlungen* 266, no. 1 (2012): 39–50.

15 Peg Yacobucci, Phone interview with the author, April 1, 2016.

16 Sessa, Skype interview with the author, January 21, 2016.

17 Adolf Naef, *Die fossilen Tintenfische: Eine paläozoologische Monographie* (Fischer, 1922).

18 Z. Lewy, "Octopods: Nude Ammonoids That Survived the Cretaceous–Tertiary Boundary Mass Extinction," *Geology* 24, no. 7 (1996): 627–630.

19 The naming of the argonaut is actually even more intricate. The label *Argonauta* was first given to the empty shells alone, and for a long time the octopus "nautiluses" that were occasionally found inside the shell "argonauts" were thought to be living like hermit crabs in a borrowed shell.

20 Julian K. Finn and Mark D. Norman, "The Argonaut Shell: Gas-Mediated Buoyancy Control in a Pelagic Octopus," *Proceedings of the Royal Society of London B: Biological Sciences* (2010): rspb20100155.

21 Lewy, "Octopods: Nude Ammonoids."

22 Roger A. Hewitt and Gerd E. G. Westermann, "Recurrences of Hypotheses about Ammonites and Argonauta," *Journal of Paleontology* 77, no. 4 (2003): 792–795.

23 Of course, to test this hypothesis we'd need to know the actual efficiency of the shells. With this aim, Ritterbush has been 3D-printing argonaut shells along with her ammonoids.

24 Alexander Arkhipkin, Skype interview with the author, March 14, 2016.

25 Alexander I. Arkhipkin, Vyacheslav A. Bizikov, and Dirk Fuchs, "Vestigial Phragmocone in the Gladius Points to a Deepwater Origin of Squid (Mollusca: Cephalopoda)," *Deep Sea Research Part I: Oceanographic Research Papers* 61 (2012): 109–122.

26 Arkhipkin, Skype interview with the author, March 14, 2016.

27 Dirk Fuchs, Yasuhiro Iba, Christina Ifrim, et al., *Longibelus* gen. nov., a New Cretaceous Coleoid Genus Linking Belemnoidea and Early Decabrachia," *Palaeontology* 56, no. 5 (2013): 1081–1106.

7. Reinvasion

1 Francesca A. McInerney and Scott L. Wing, "The Paleocene-Eocene Thermal Maximum: A Perturbation of Carbon Cycle, Climate, and Biosphere with Implications for the Future," *Annual Review of Earth and Planetary Sciences* 39 (2011): 489–516.

2 Henk Brinkhuis, Stefan Schouten, Margaret E. Collinson, et al., "Episodic Fresh Surface Waters in the Eocene Arctic Ocean," *Nature* 441, no. 7093 (2006): 606–609.

3 Ellen Thomas, "Descent into the Icehouse," *Geology* 36, no. 2 (2008): 191–192.

4 Andrew Packard, "Cephalopods and Fish: The Limits of Convergence," *Biological Reviews* 47, no. 2 (1972): 241–307.

5 J. A. Mather and R. C. Anderson, "What Behavior Can We Expect of Octopuses?" *The Cephalopod Page*, http://www.thecephalopodpage .org/behavior.php (accessed January 28, 2017).

6 R. A. Byrne, U. Griebel, J. B. Wood, and J. A. Mather, "Squid Say It with Skin: A Graphic Model for Skin Displays in Caribbean Reef Squid (*Sepioteuthis sepioidea*)," in "Proceedings of the International Symposium 'Coleoid Cephalopods Through Time,' 17–19 September 2002," ed. K. Warnke, H. Keupp, and S. Boletzky, *Berliner paläobiologische Abhandlungen* 3 (2003): 29–35.

7   Neale Monks, "A Broad Brush History of the Cephalopoda," *The Cephalopod Page*, http://www.thecephalopodpage.org/evolution.php (accessed 20 January 20, 2017).

8   The Smithsonian National Museum of Natural History has an excellent animated presentation about the evolution of whales: https://ocean.si .edu/ocean-videos/evolution-whales-animation (accessed January 28, 2017).

9   Neale Monks, "Tertiary Cephalopods or Where Did All the Ammonites Go?" *Deposits Magazine*, November 8, 2016, 28–31.

10 David R. Lindberg and Nicholas D. Pyenson, "Things That Go Bump in the Night: Evolutionary Interactions Between Cephalopods and Cetaceans in the Tertiary," *Lethaia* 40, no. 4 (2007): 335–343.

11 Dirk Fuchs, Skype interview with the author, January 18, 2016.

12 Jakob Vinther, Skype interview with the author, March 15, 2016.

13 Thomas Clements, Caitlin Colleary, Kenneth De Baets, and Jakob Vinther, "Buoyancy Mechanisms Limit Preservation of Coleoid Cephalopod Soft Tissues in Mesozoic *Lagerstätten*," *Palaeontology* 60, no. 1 (2017): 1–14.

14 Ammonia was first isolated from ammonia salt, in fact, which was found in deposits in Egypt near the temple of Ammon — the ram's horn God that gave ammonoids their name. Hence, ammonia and ammonoid have the same linguistic root, and it is even possible that ammonoids sequestered ammonia in their tissue like squid.

15 I would like to take off every hat I own to Thomas Clements and collaborators, who endured six weeks of decay experiments for the sake of science. "By Day 9, the water around the octopus carcass began to discolour and turn black, accompanied by an obnoxious smell," they dutifully report in "Buoyancy Mechanisms." "The water within the jars with the octopus samples turned a milky red colour, while the liquid accompanying the squid tissues turned milky white with a thick yellow immiscible scum layer at the surface that had a distinctive sickly sweet smell" (8). May they never again be afflicted with such a "distinctive" experience.

16 Vinther, Skype interview with the author, March 15, 2016.

17 Vinther, Skype interview with the author, March 15, 2016.

18 Robyn Crook and Jennifer Basil, "A Biphasic Memory Curve in the Chambered Nautilus, *Nautilus pompilius* L. (Cephalopoda: Nautiloidea)," *Journal of Experimental Biology* 211, no. 12 (2008): 1992–1998.

19 Peter Ward, Frederick Dooley, and Gregory Jeff Barord, "Nautilus: Biol-
    ogy, Systematics, and Paleobiology as Viewed from 2015," *Swiss Journal
    of Palaeontology* 135, no. 1 (2016): 169–185.
20 Claire Régnier, Guillaume Achaz, Amaury Lambert, et al., "Mass Ex-
    tinction in Poorly Known Taxa," *Proceedings of the National Academy of
    Sciences* 112, no. 25 (2015): 7761–7766.
21 Jocelyn Sessa, Skype interview with the author, January 21, 2016.
22 Mariette Le Roux, "Scientists Warn of 'Deadly Trio' Risk to Ailing
    Oceans," October 3, 2013, https://phys.org/news/2013-10-scientists
    -deadly-trio-ailing-oceans.html (accessed January 28, 2017).

8. Where Are They Now?

1 Kenneth De Baets, Skype interview with the author, July 24, 2015.
2 Helen Scales, *Spirals in Time: The Secret Life and Curious Afterlife of Seashells* (Bloomsbury Sigma, 2015).
3 The "Tree Octopus" has got to be one of the oldest websites around, and
   it really merits a look: http://zapatopi.net/treeoctopus/ (accessed January 28, 2017).
4 "Octopus Walks on Land at Fitzgerald Marine Reserve," uploaded by
   tuantube, with more than 7 million views. Everyone loves a friendly octopus! https://www.youtube.com/watch?v=FjQr3lRACPI (accessed
   January 28, 2017).
5 I spent way too long figuring out exactly what percentage of the earth's
   surface is covered by freshwater. I started with these facts: 0.01 percent
   of all water on Earth is fresh surface water, and 87 percent of that is in
   lakes. I hied me to Wikipedia's list of lakes by area, which reels off the
   thirty-eight lakes with a surface area greater than 4,000 square kilometers. Eight of these are saline, including the Caspian Sea (did you know
   that's considered a lake? I didn't either!), which leaves thirty freshwater lakes (a suspiciously round number, but okay), whose surface area
   sums up to 616,422 square kilometers. If that's 87 percent of fresh surface water, then 100 percent of fresh surface water would be that number divided by 0.87, which is 708,531 square kilometers. But we must
   remember these are only the *largest* lakes, so instead of comprising 87
   percent of fresh surface water, let's say they're only 80 percent. This gets
   us to roughly 770,000 square kilometers of the earth's surface covered

by freshwater. Divide this by the total surface area of the planet (510.1 million square kilometers) and we get 0.15 percent of the earth's surface area covered with freshwater. Boom!

6 Clyde F. E. Roper and Elizabeth K. Shea, "Unanswered Questions about the Giant Squid *Architeuthis* (Architeuthidae) Illustrate Our Incomplete Knowledge of Coleoid Cephalopods," *American Malacological Bulletin* 31, no. 1 (2013): 109–122.

7 For a funny and fascinating takedown of exaggerated squid sizes, see "Whale Sharks and Giant Squids: Big or Bu!!$hit?" by Craig McClain, assistant director of science for the National Evolutionary Synthesis Center and chief editor of *Deep Sea News*: https://www.deepseanews .com/2013/02/whale-sharks-and-giant-squids-big-or-buhit (accessed January 28, 2017).

8 Danna J. Staaf, William F. Gilly, and Mark W. Denny, "Aperture Effects in Squid Jet Propulsion," *Journal of Experimental Biology* 217, no. 9 (2014): 1588–1600. The first draft of this paper was titled "The Littlest Squid." Didn't make the cut.

9 Norbert Cyran, Lisa Klinger, Robyn Scott, et al., *Characterization of the Adhesive Systems in Cephalopods* (Springer, 2010).

10 Julian K. Finn, Tom Tregenza, and Mark D. Norman, "Defensive Tool Use in a Coconut-Carrying Octopus," *Current Biology* 19, no. 23 (2009): R1069–R1070.

11 Christine L. Huffard, Farnis Boneka, and Robert J. Full, "Underwater Bipedal Locomotion by Octopuses in Disguise," *Science* 307, no. 5717 (2005): 1927–1927.

12 Hendrik J. T. Hoving and Bruce H. Robison, "Vampire Squid: Detritivores in the Oxygen Minimum Zone," *Proceedings of the Royal Society of London B: Biological Sciences* (2012): 4559–4567.

13 Julia S. Stewart, Elliott L. Hazen, Steven J. Bograd, et al., "Combined Climate- and Prey-Mediated Range Expansion of Humboldt Squid (*Dosidicus gigas*), a Large Marine Predator in the California Current System," *Global Change Biology* 20, no. 6 (2014): 1832–1843.

14 Roy Caldwell, "Death in a Pretty Package: The Blue-Ringed Octopuses," *Freshwater and Marine Aquarium Magazine*, 23, no. 3 (2000): 8–18; reprinted http://www.thecephalopodpage.org/blueringi.php (accessed January 28, 2017).

15 Roy Caldwell and Christopher D. Shaw, "Mimic Octopuses: Will We Love Them to Death?" *The Cephalopod Page*, http://www.thecephalopod page.org/mimic.php (accessed January 28, 2017).

16 Ibid.

17 Jacquet, Jennifer, Becca Franks, Peter Godfrey-Smith, and Walter Sánchez-Suárez. "The case against octopus farming." *Issues in Science and Technology* 35, no. 2 (2019): 37-44.

18 The latest information about the California market squid fishery can be found on the California Department of Fish and Wildlife website: https://www.wildlife.ca.gov/Conservation/Marine/CPS-HMS /Market-Squid (accessed January 28, 2017).

19 Hoving, Henk-Jan T., S. L. Bush, S. H. D. Haddock, and B. H. Robison. "Bathyal feasting: post-spawning squid as a source of carbon for deep-sea benthic communities." *Proceedings of the Royal Society B: Biological Sciences* 284, no. 1869 (2017): 2017-2096.

20 Marnie Hanel, "The Octopus That Almost Ate Seattle," *New York Times*, October 16, 2013.

21 The two octopus species on the IUCN Red List are *Opisthoteuthis mero* (endangered) and *Opisthoteuthis chathamensis* (critically endangered). At least one other is listed as vulnerable, and many more are considered "data deficient" — that is, we just don't know enough about them to say how they're doing. http://www.iucnredlist.org/details/163144 /0; http://www.iucnredlist.org/details/162917/0 (accessed January 28, 2017).

22 Gregory J. Barord, Frederick Dooley, Andrew Dunstan, et al., "Comparative Population Assessments of *Nautilus* sp. in the Philippines, Australia, Fiji, and American Samoa Using Baited Remote Underwater Video Systems," *PloS One* 9, no. 6 (2014): e100799.

23 Gregory Barord, Email to the author, November 15, 2016.

Epilogue: Where Are They Going?

1 Zoë A. Doubleday, Thomas A. A. Prowse, Alexander Arkhipkin, et al., "Global Proliferation of Cephalopods," *Current Biology* 26, no. 10 (2016): R406–R407.

2 The subject of whether or not there is a worldwide jellyfish bloom is contentious. See Richard D. Brodeur, Jason S. Link, Brian E. Smith, et

al., "Ecological and Economic Consequences of Ignoring Jellyfish: A Plea for Increased Monitoring of Ecosystems," *Fisheries* 41, no. 11 (2016): 630–637; and Marina Sanz-Martín, Kylie A. Pitt, Robert H. Condon, et al., "Flawed Citation Practices Facilitate the Unsubstantiated Perception of a Global Trend toward Increased Jellyfish Blooms," *Global Ecology and Biogeography* 25, no. 9 (2016): 1039–1049. Finally, here's an intriguing citizen science initiative aimed at gathering more information about jelly sightings: http://www.jellywatch.org/ (accessed January 28, 2017).

3   Elizabeth Kolbert, *The Sixth Extinction: An Unnatural History* (A & C Black, 2014).

4   Jose C. Xavier, A. Louise Allcock, Yves Cherel, et al., "Future Challenges in Cephalopod Research," *Journal of the Marine Biological Association of the United Kingdom* 95, no. 5 (2015): 999–1015.

5   Eric Edsinger-Gonzales, Skype interview with the author, October 5, 2016.

6   Xavier at al., "Future Challenges," 9.

7   Ibid., 6.

8   The updated two-volume edition of the ammonoid "holy book" that was fortuitously published just as I was researching this book: Christian Klug, Dieter Korn, Kenneth De Baets, et al., eds., *Ammonoid Paleobiology: From Anatomy to Ecology*, Topics in Geobiology, vol. 43 (Springer, 2015); Christian Klug, Dieter Korn, Kenneth De Baets, et al., eds., *Ammonoid Paleobiology: From Macroevolution to Paleogeography*, Topics in Geobiology, vol. 44 (Springer, 2015).

9   Isabelle Kruta, Skype interview with the author, March 31, 2016

10 Isabelle Kruta, Isabelle Rouget, Sylvain Charbonnier, et al., "*Proteroctopus ribeti* in Coleoid Evolution," *Palaeontology* 59, no. 6 (2016): 767–773.

Index

Page numbers in *italics* indicate illustrations.

About the Author

DANNA STAAF earned a PhD in invertebrate biology from Stanford University and has been studying cephalopods for decades. She is also the author of *Nursery Earth: The Wondrous Lives of Baby Animals and the Extraordinary Ways They Shape Our World*. Her writing on marine life has appeared in *Science*, *Atlas Obscura*, and many other outlets, while her research has appeared in the *Journal of Experimental Biology*, *Aquaculture*, and others, as well as in numerous textbooks. She lives with her family in Northern California.

dannastaaf.com | 🐦dannastaaf | 📷dannajoystaaf | 📘bananacough